— THE BOOK OF —
PROPOSITION
BETS

Using Mathematics to Reveal the Odds
of Friendly (and Not-So-Friendly) Wagers

OWEN O'SHEA

 Prometheus Books
Guilford, Connecticut

(PB) Prometheus Books

An imprint of The Rowman & Littlefield Publishing Group, Inc.
4501 Forbes Blvd., Ste. 200
Lanham, MD 20706
www.rowman.com

Distributed by NATIONAL BOOK NETWORK

British Library Cataloguing in Publication Information Available

Library of Congress Cataloging-in-Publication Data

Name: O'Shea, Owen, 1956–, author.
Title: The book of proposition bets : using mathematics to reveal the odds of
   friendly (and not-so-friendly) wagers / Owen O'Shea.
Description: Lanham, MD : Prometheus, [2021] | Includes bibliographical
   references and index.
Identifiers: LCCN 2021000797 (print) | LCCN 2021000798 (ebook) |
   ISBN 9781633886742 (paperback) | ISBN 9781633886759 (ebook)
Subjects: LCSH: Gambling systems. | Chance. | Probabilities.
Classification: LCC GV1302 .O84 2021 (print) | LCC GV1302 (ebook) |
   DDC 795—dc23
LC record available at https://lccn.loc.gov/2021000797
LC ebook record available at https://lccn.loc.gov/2021000798

∞™ The paper used in this publication meets the minimum requirements of American National Standard for Information Sciences—Permanence of Paper for Printed Library Materials, ANSI/NISO Z39.48-1992

To Des MacHale and Donal Hurley,
for their encouragement and support over many years.

CONTENTS

ACKNOWLEDGMENTS

I want to thank Connor Stoyle for drawing the illustrations in this book. I would also like to thank Connor for the computer programs he has written to confirm specific results in probability theory used in this work. Thanks to the editorial staff at Prometheus Books, particularly Kellie Hagan, for their very helpful suggestions and for their efficient and diligent work in preparing and editing this book. Finally, thanks to Drs. Lara Hawchar and Cónall Kelly, both from the School of Mathematical Sciences, University College, Cork, Ireland, who ran Monte Carlo simulations on their computers to obtain results that assisted me in compiling specific information in this book.

INTRODUCTION

In the modern world, the theory of probability is used extensively in mathematics, science, engineering, medicine, and, of course, gambling. The use of probability theory to deceive members of the public with proposition bets is the subject of this book.

Individuals offering the proposition bet are often referred to with different names, but *grifter, hustler, fraudster,* and *con artist* are prevalent. *Con artists* try to win your *con*fidence (thus the name). They offer what *appears* to be a genuine opportunity or genuine gamble that may lead you to winning cash or some other valuable item.

Marks (sometimes referred to as *suckers* by the gambling fraternity) are easily deceived or tricked by con artists. The origin of the term *mark* is interesting. In the distant past, when a member of the public was swindled when engaging in a rigged carnival game, the fraudster would place some powder on the palm of their hand and then place their hand on the back of the victim, seeming to commiserate with them on their bad luck at not winning at the game. Of course, the powdered hand leaves a mark on the back of the unfortunate victim, which helps identify them to other unscrupulous con artists operating in the carnival. These other grifters then attempt to further take advantage of the unsuspecting dupe. In this way the victim became a *marked* person. From then on, the term *mark* was used to describe the victim of a fraud or proposition bet.

A proposition bet, or a *sucker bet* as it is sometimes called, is when an individual makes an apparently attractive bet to someone (the mark) who is hopefully (from the hustler's point of view) easily deceived.

The bet appears so attractive to the mark that the sucker erroneously believes that the odds of winning the bet are in their favor. The fraudster wants suckers to think that they (the marks) are onto a good thing. The swindler lays this bait in order to trap the unwary.

This is not to suggest that all those who are on the receiving end of a sucker bet are easily deceived. Far from it! Some very intelligent people have been lured into accepting sucker bets because the wagers so cleverly conceal the true probabilities behind them.

One of the most famous proposition bets is based on what is known as the Monty Hall Paradox. The bet is simple to state. The hustler places 2 red queens and a black ace facedown on a table. You are told by the hustler that 2 of the cards are red queens and 1 of them is a black ace. The con artist also tells you that they know where the black ace is located among the three cards on the table. You do not, of course, know how the 3 cards are distributed.

Let the position of the cards be designated A, B, and C. Let us assume that the card at position B is the black ace. You do not know this, but the hustler does. The hustler now bets you $10 at even-money odds that you cannot point to the black ace. The probability you will correctly do this is ⅓. Suppose you point to position A as the card you believe is the black ace. The hustler now turns over 1 of the other 2 cards (let us assume, in this example, the card at position C) and reveals it to be a red queen. Then the hustler removes that card from the table.

There are now only 2 cards on the table: A and B. One of the cards, A, is the card you chose. The hustler asks you if you want to change your mind and choose the other facedown card, at position B, on the table.

If you decide to stick with your original choice you will win this bet about 1 time in every 3, because the probability of finding the black ace is ⅓. The chances of your losing the bet are therefore ⅔. Thus, if you decide to stick with your initial choice the odds favor the hustler winning the bet.

However, if you decide to switch from your initial choice, A, to the other unturned card, the chances of your winning rises to ⅔, and the chance of you losing is ⅓.

Most people, on being given this choice, believe it does not matter whether one sticks with the initial choice or whether one changes one's mind and switches to the other unturned card. However, it does matter!

Here's the thing. If you do not switch, your chance of picking the black ace is ⅓, so think of the other unturned card as the "winning card" with probability of ⅔. Therefore, if you switch ⅔ of the time, you switch to the unturned "winning card." Consequently, by switching you double your chances from ⅓ to ⅔ of picking the black ace.

Suppose a con artist is regularly offering this bet to various marks at various locations. At a bet of say, $10 a round, when the mark wins, they win $10. About ⅓ of the time, the mark will choose the wrong card. If the mark decides not to switch from their original choice, they lose. This will occur about ⅓ of the time. But the hustler wins about ⅔ of the time and therefore for every $10 the mark wins, the hustler wins $20. Therefore, the con artist is winning this bet ⅔ of the time and in so doing, is making a tidy profit.

The true probability behind this proposition bet is so cleverly hidden that one of the greatest mathematicians who ever lived, Paul Erdős (1913–1996), was fooled by it. So, don't feel too bad if some of the proposition bets in this book fool you. If they do, you're in good company!

So how likely is it that you will be offered a sucker bet? That depends on your lifestyle and I suppose, to some degree, whether you are a betting person. But even if you are not, and you frequently socialize with others, it is possible—probably likely—that at some point you will be offered a sucker bet. Of course, you will not be informed that it is a sucker bet! No, that is not the way the swindler operates. The sucker bet will arrive attractively packaged and disguised to lure you into accepting it.

Depending on how many dollars are resting on the wager, the outcome—if you win the bet—could leave you a very happy camper. But alas, the structure of the proposition bet, coupled with the theory of probability, is against this happening. It is more likely that you will lose the bet. That is why sucker bets are sprung on innocent folk! If your financial loss is small, look upon your loss as the purchasing price of learning something new. If, however, you were to endure a heavy loss . . . well that is when you begin to realize that perhaps you should have purchased this book.

Thus as you go through the winding road of life, you can take it for granted that one evening, when you least expect it, a person will walk up to you and say something like this: "Listen pal, as you well know there are 12 months in a year. Now I am a betting dude. I like to wager now and then. I guess it adds a little excitement to my life. So, you know what? I am going to offer you a little bet and a handy opportunity for you to make $10.

"I am going to ask you to secretly write down on a piece of paper the names of just 3 different months. That's all! I will then take 3 guesses, and I mean just 3 guesses. No fishing or any of that sort of thing. No, my friend, I don't go in for such capers. Everything is on the straight and level with me! That is the way I operate. I will bet you $10 at even money odds that if I take 3 guesses, I will correctly guess at least one of the months you have written down. Now look here, I am putting $10 down on this nearby table. I am asking you to match it. Thus, if one of my guesses is correct, I will collect the 20 bucks on the table. However, if you win the bet *you* take the $20. Now the odds must favor you, my friend, because I am allowed only 3 guesses, and every calendar wizard on this side of the Mississippi knows that there are 12 different months in a year. So come on, what are you waiting for? Don't you want the chance to make a handy 10 bucks?"

Of course, while the con artist is stating all this the thought is racing through your mind whether you should accept the bet. Having thought about it for a minute or two it seems to you that the odds are indeed in your favor. The stranger is allowed only 3 guesses, and there are, as they have said, 12 months in a year. That means they have 1 chance in 4 of guessing right, doesn't it? Surely the odds in this bet favor you.

It is this type of reasoning that may make you inclined to accept the wager and hopefully make a profit of $10 for yourself.

Of course, that is precisely the type of reasoning that the stranger is hoping you will adopt. You see, the stranger that is offering this bet to you is probably more accurately described as a hustler. They are offering a bet to you in which the odds apparently favor you. However, the odds favor them. Big time!

So how are the true odds in this bet calculated? It is not too difficult to do the math, but the odds are cleverly concealed in offering the proposition bet by the con artist's patter. They make it appear that the bet is overwhelmingly in your favor, while the odds are in favor of the con artist.

Here's the way to calculate the true odds in this proposition bet: there are 12 months in a year. You have secretly written down the names of 3 months. Only you know what you have written down. But the hustler doesn't need to know this information. The laws of probability are enough to make the wager favorable to them.

To calculate the odds, we first calculate the probability that the con artist does *not* correctly guess the names of any of the 3 months you have written down. The probability that they will *not* correctly guess the first month you have chosen is $\frac{9}{12}$. (Recall that they cannot name any of the 3 months you have written down if they are to incorrectly guess the names of the 3 months you chose.) The probability that they will *not* correctly guess the second month you have written down is $\frac{8}{11}$. The probability that they will *not* correctly guess the third month you have selected is $\frac{7}{10}$.

These 3 fractions must then be multiplied together to find the probability that the con artist will *not* correctly guess any of the 3 months you have written down.

$$\frac{9}{12} \times \frac{8}{11} \times \frac{7}{10} = \frac{504}{1320} = 0.3818 +$$

Thus, the probability that the con artist will *not* correctly guess any of the names of the 3 months you have written down is 0.3818+. If we subtract this from 1 (*certainty*) we will obtain the probability that the hustler WILL correctly guess at least one of the names of the 3 months you have chosen.

$$1 - 0.3818+ = 0.6181+$$

Therefore, the probability that the con artist will win the bet is 0.6181. In percentage terms this equals 61.81 percent.

The odds therefore are approximately 60 to 40, which simplify to 3 to 2, in favor of the con artist winning this wager.

Thus, over the long term, in every 5 bets, the con artist can expect to win 3 of them. The swindler therefore is onto a good thing.

These types of bets are known to gamblers throughout the world as *proposition bets*, or even more famously, as *sucker bets*. They are offered in the hope that the perceived *sucker* will accept the bet and will—more likely than not—lose the wager and, in the process, yield a nice profit for the hustler.

Of course, the hustler's individual bets in financial terms may not, initially, seem very large. They may range from $5 to $50, and on rare occasions may even be a lot more. However, the swindler wins big time because they are regularly offering—and winning often—the bets. If they offer a bet where the true odds that they will win are 7 out of 10, then in the long term they will win that bet 7 times out of 10. In that way their accumulated winnings can become a tidy sum in a relatively short period.

It is my experience that these proposition bets appeal to many, even to those who have no interest in gambling generally. Most folk are astonished when the true odds to the various proposition bets given in the book are revealed and explained.

I give the sources to the various sucker bets when I am aware of where they have appeared in print. I will point out here that I had come across many of these proposition bets in the days of my youth when I gambled for a few years at games of cards and dice. I took note of the sucker bets as I learned them from underground gamblers and have kept them in my files over the years.

The source for many of these sucker bets is found in gambler's folklore because that is where I learned of them. I played against and associated with various underground gamblers when I was a young man. Having consulted with some of these underground gamblers, they have informed me that they do not wish to be named. I will of course respect their wishes.

This book does not attempt to glamorize gambling per se. On the contrary! By illustrating the odds in these proposition bets, the book attempts to forewarn members of the public not to accept these wagers if they are offered to them. Remember, to be forewarned is to be forearmed! The book is written for educational and entertainment purposes only. If you decide to use ideas contained in this book, you will take full responsibility for your actions.

The book you are now holding contains more than 50 of these mathematical sucker bets. They mainly involve proposition bets relating to playing cards and dice.

In the second half of the book I give a brief summary of several classic proposition bets that are designed to part you from your hard-earned cash. At the end of the book I include a few more mathematical proposition bets.

The reader is not required to have any knowledge of probability theory to understand the mathematics behind these cunning wagers. If you can understand basic arithmetic, you will be able to follow the mathematics behind most, if not all, of these intriguing propositions.

What can you expect from reading this book? Hopefully, for the relatively small asking price to purchase the book you will learn the basic lesson in life (that many have not apparently yet learned!) that if something appears too good to be true, then it probably is! That fact is true of gambling propositions also.

But you will also learn more. You will learn the basic truth lurking behind all mathematical sucker bets: that the laws of probability can be highly deceptive and that outcomes or events that appear highly improbable are—quite often—very probable.

Hopefully along the way you will also enjoy (and perhaps appreciate) the clever thinking that lay behind many of these wagers.

I suggest you buy this book and if you do, I think there is a high probability that you will enjoy it!

THE 3 HALF-DOLLARS BET

"It is likely that unlikely things will happen."—*Aristotle*

PROBLEM

A stranger may approach you in a bar one night.

The stranger gets talking to you and tells you that since childhood they have been interested in gambling. They say that they would like to offer a little wager and are prepared to pay generous odds to you should you win in a little gambling experiment that they have devised.

You inquire as to what the experiment is.

The stranger says that they will ask you to give 3 half-dollar coins to one of the customers (of your choosing) in the bar. That person is to then secretly toss the 3 half-dollars simultaneously in the air. They are then to place a drink coaster over each half-dollar so that no one knows what faces are shown on the 3 coins.

The hustler then says something like the following to you: "I have no idea what faces are showing on the 3 half-dollars. I do know however, that 2 of the 3 half-dollars necessarily have the same face showing. Now consider the third half-dollar. The face showing on that coin is either the same face as those showing on the other 2 half-dollars, or it is not the same. Therefore, the probability that all 3 coins have the same face showing must be 50 percent."

The hustler continues. "I am a fair betting person. As I have just demonstrated it is even odds that the third coin is showing the same face as the other 2 half-dollars. Therefore, I am going to bet you $10 that this is *not* the case. In other words, I am going to bet you that the 3 half-dollars that have been tossed do *not* show the same face. Also, because I am generous, I am going to give you odds of 2 to 1 in your favor. What can be fairer than that? Now, my friend, I am giving you a genuine opportunity of making a tidy profit of $20. Do you want to take that opportunity? If you do, then put your $10 down on top of my $10 here on this table. The winner of the bet will take all $20. What could be fairer?"

You are there in the bar listening to all this. The bet does seem very tempting, especially since the stranger is offering odds of 2 to 1 to you should you accept the bet and win.

You are contemplating accepting the bet.

The question is this: Should you accept this wager?

SOLUTION

No way!

The stranger is a hustler. The true odds of your winning this bet are 1 in 4. If this bet is repeated many times the con artist will win the bet 3 out of every 4 times.

Look at it this way. The 2 sides of the half-dollar depict the profile of President John F. Kennedy on the obverse side (face A) and the seal of the president of the United States on the reverse side (face B).

When the 3 coins are tossed the following 4 outcomes are possible:

- There is 1 way that the 3 half-dollars have face A showing.
- There are 3 ways that 2 of the 3 half-dollars have face A showing.
- There are 3 ways that 1 of the 3 half-dollars have the face A showing.
- There is 1 way that none of the three half-dollars have face A showing.

We see that there are 8 different possibilities in which the 3 coins could fall. In 2 of these ways, the 3 coins show the same face. In the other 6 ways, the 3 coins do not have the same face showing. Therefore, the true odds are 6 to 2, which reduce to 3 to 1, that the hustler will win the bet.

Saying the same thing another way, you have 1 chance in 4 of winning the bet. The con artist has 3 chances in 4 of winning the wager.

Thus, even if the hustler gives you odds of 2 to 1, the bet in the long run still favors him. Suppose the hustler does the bet 120 times, at $10 per bet. The con artist will win the bet about 90 times, winning $900. The 30 times they lose, they will pay out 30 times $20, or $600. The swindler has still made a profit of $300. That equates to an *average* profit per game of $2.50 for the con artist.

That is a nice profit margin for the swindler.

Thus, if you are offered this bet, it is best to decline it.

CHOOSING 3 RANDOM CARDS WITH NO SPADE

"I have no certainties, at most probabilities."—*Renato Caccioppoli*

PROBLEM

The proposition is that you will choose 3 random cards from a 52-card deck and that there will not be *exactly* 1 spade among those 3 cards.

You may be at a social gathering one evening when a stranger (who also happens to be a con artist) walks up to you and offers the following proposition bet: "Take a shuffled deck of 52 cards and turn the deck facedown. Now select any 3 cards one at a time from the deck. I am prepared to bet you $5—at odds of even money—that there is not *exactly* 1 spade among those 3 cards."

The con artist may spice up the conversation by adding the following lines: "There are 13 spades in the deck of 52 cards. Therefore, if one chooses just 1 card at random, the chance it is a spade is 1 in 4. However, I am giving you the opportunity of selecting 3 cards at random. Therefore, there must be 3 chances in 4 that at least 1 spade is among the 3 cards you pick. Thus, it is likely that there is 1 spade among the 3 cards you choose. Thus, the bet favors you, my friend. But I am in a generous mood today, and that is why I am giving such generous odds. So maybe today is your lucky day."

Of course, this kind of talk from the stranger may lead you to reason that 1 of the 3 cards is probably a spade, and therefore you accept the bet.

What is the probability that *exactly* 1 of the cards is a spade?"

SOLUTION

This hustle is an old favorite with swindlers.

Let's look a little more closely at the proposition. You are asked to draw 3 cards from a deck of 52 cards. What are the odds that *exactly* 1 of the cards is a spade?

There are 13 spades in a deck of 52 cards. The probability that the first card selected is a spade is therefore $^{13}\!/_{52}$. The probability that the second card selected is *not* a spade is $^{39}\!/_{51}$, because there are 39 non-spades in the deck. The probability that the third card selected is *not* a spade is $^{38}\!/_{50}$, because there are still 38 non-spades in the deck.

Multiply these 3 probabilities together and one obtains 0.145294+. In percentage terms this equals 14.5294+. This is the probability that the first selected card is a spade and the other 2 are not.

However, any 1 of the 3 selected cards can be the spade. Therefore, we multiply 14.5294+ by 3 to obtain 43.5882+.

Thus, the probability that *exactly* 1 of the selected 3 cards is a spade is 43.58823 percent.

Therefore, the probability that the selected 3 cards do *not* contain *exactly* 1 spade is 1 − 0.435882+. This equals 0.564118+. In percentage terms this equals 56.4118.

Thus, it is more likely that there will not be *exactly* 1 spade among the 3 cards.

The con artist can expect to win this bet about 56 times in every 100 wagers that they make. If the con artist plays 100 such games at $5 a game, they can expect to win $280 and lose $220. That means the hustler makes a tidy profit of about $60 dollars for every $500 waged. That equals a 12 percent profit margin, which is not a bad return for the swindler.

The best advice is to stay away from this bet, as it favors the hustler.

FIRST 3 FACE CARDS BEING THE SAME SUIT

"Statistically the probability of any one of us being here is so small that you would think the mere possibility of existence would keep us all in a contented dazzlement of surprise."—*Lewis Thomas*

PROBLEM*

The proposition is that of the first 3 different face cards (king, queen, jack, in any order) you come across in dealing 1 card at a time from a shuffled deck, 2 of those 3 face cards will be of the same suit.

One evening when you are with some close friends you may be approached by a friendly stranger.

Within a short period of time the stranger will steer the conversation around to gambling and offer a proposition bet that will appear attractive to you.

The stranger will produce a deck of cards and ask you to examine the cards to verify that it is a normal deck of 52 cards. The con artist will then ask you to thoroughly shuffle the deck.

You do so, probably wondering what is coming next.

The con artist will then ask you to deal from the facedown deck 1 card at a time and to turn each card dealt faceup and says they are prepared to bet $10 at even money odds that of the first 3 different face cards that are dealt (the king, queen, or jack in any order), 2 of those cards will be of the same suit.

The hustler may expand a little at this point to explain exactly what the bet entails. The swindler will emphasize to you that it does not matter in what order the 3 face cards are dealt. All that matters, they will say, is that the first king, queen, or jack that you deal, in whatever order, be observed. The con artist will state that they are betting $10 at even money odds that 2 of those 3 cards will be of the same suit.

You may well contemplate the bet. You are fully aware that in a normal deck there are 4 kings, all of a different suit; 4 queens, all of a different suit; and 4 jacks, all of a different suit. Thus, you may well conclude that the probability is small that 2 of the first 3 cards, king, queen, or jack in whatever order that they may arise, are of the same suit. Consequently, you may be inclined to accept the wager.

The question is this: Should you accept the bet?

---

*Source: [Author unknown], *The Complete Home Entertainer* (London: Odhams Books, c. 1940).

SOLUTION

You should not accept this wager. The odds that the con artist will win this bet are 5 to 3 in their favor.

To find how these odds are calculated let us examine the number of ways we can choose 3 different face cards so that *none* of them are of the same suit.

There are 4 kings, 4 queens, and 4 jacks in a normal deck. They can be arranged or dealt in 4 × 4 × 4 or 64 different ways. The order in which the 3 different face cards are arranged in a shuffled deck does not alter the odds in any way.

Let us assume the jack is the first card one comes across as one scans the deck. It can be any 1 of the 4 jacks. Thus, there are 4 ways the jack can be selected.

Let us assume the second card one comes across is the queen. If the queen is to be of a different suit than the jack just dealt, the queen can only be 1 of 3 suits.

Let us assume the king is the third face card we encounter in the deck. If the king is to be a different suit from the jack and queen just dealt, the king can only be 1 of 2 suits.

Therefore, the number of ways that 3 face cards can be encountered in a shuffled deck so that they are all a different suit is 4 × 3 × 2 or 24.

We calculated earlier that there are 64 ways of arranging the 12 face cards. Therefore, there are 64 − 24 or 40 ways that the 12 face cards can be encountered in a shuffled deck where at least 2 of the 3 face cards are of the same suit.

Thus, the probability that 2 of the 3 different face cards are of the same suit is $^{40}/_{64}$. This fraction simplifies to $^{5}/_{8}$.

Therefore, in the long term, the con artist will win this wager 5 times out of every 8 bets and will lose only 3 such wagers.

Thus, the odds are 5 to 3 in favor of the hustler.

4

THE 4 ACES AND 2 KINGS PROPOSITION

"The million, million, million . . . to one chance happens once in a million, million, million . . . times no matter how surprised we may be that it results in us."—*Ronald Aylmer Fisher*

PROBLEM*

As sure as night follows day, a con artist will walk into a bar in your locality one evening and pick a mark out of the crowd. The grifter will make some small talk and will eventually offer the following proposal to the mark: they will ask the mark to take 4 aces and 2 kings from a deck of 52 cards. The hustler will then ask the mark to shuffle the 6 cards and to place them facedown on a table.

The con artist will then offer the following proposition to the mark: they will ask the mark to choose any 2 cards from the 6 cards on the table and then turn them over. If that person succeeds in choosing 2 aces, they win $20. It is as simple as that! However, if the 2 cards they choose are not 2 aces they lose $20. In other words, the con artist bets the mark 20 bucks at even money that they cannot pick 2 aces from the 6 facedown cards.

The swindler will place a $20 bill on the table and ask the mark to "cover" it (that is, for the mark to place 20 bucks on the table also). The winner of the bet collects all $40.

The successful outcome of the bet appears to be quite easy to achieve. After all, there are 4 aces among the 6 cards. The mark may believe the chances are 4 out of 6, which reduces to 2 out of 3 that they will choose 2 aces. The mark believes that the odds in this little wager are very much in their favor.

Of course, that is precisely what the con artist wants the mark to believe.

The question is this: Should the mark accept this bet?

---

*Source: Martin Gardner, *Entertaining Mathematical Puzzles* (New York: Dover, 1961), 71 and 72.

SOLUTION

No, the mark should not accept this bet.

Suppose that the 6 cards on the table consist of the 4 aces, plus the king of clubs and the king of spades.

To clearly see the reason why this is a bad bet to accept, partition the 6 cards into 2 groups as follows:

1. King of clubs and king of spades
2. Ace of clubs, ace of spades, ace of hearts, and ace of diamonds

Let us suppose the mark chooses the king of clubs as their first card. The second card they turn over can be any 1 of the 4 aces. That accounts for 4 different hands of 2 cards, where 1 of the 2 cards is a king.

Suppose however the mark chooses the king of spades as their first card. The second card they turn over can also be any 1 of the 4 aces. This accounts for another 4 different hands of 2 cards, where once again 1 of the 2 cards is a king.

This is a total of 8 hands of 2-card hands, where 1 of the 2 cards is a king.

Of course, the mark could also turn over the 2 kings on 2 consecutive picks. This brings the number of 2-card hands to 9, in which *at least* 1 of the 2 cards is a king.

Now let us switch our attention to the 4 aces. There are 4 ways of choosing the first ace and 3 ways of choosing the second ace. That is, there are 12 ways of choosing the 2 aces if the order of their choice is relevant. For example, one might choose the ace of spades and then the ace of hearts.

However, this is considered the same as picking the ace of hearts and then the ace of spades. Thus, the *order* of picking the 2 aces is irrelevant. If their order is irrelevant, we must divide 12 by 2, and obtain 6. Thus, there are 6 ways to choose any 2 aces from the 4 aces. This result is usually written mathematically as 4C2, which equals 6. (The expression 4C2 is read as the number of combinations of 4 objects, from which 2 objects are chosen at a time.)

Now we calculate the number of ways of choosing 2 cards from 6. There is a total of 6C2 or 15 ways of doing this.

However, we found that there are only 6 ways that 2 aces can be picked from 4 aces.

Consequently, there are (as we found earlier) 9 ways of choosing 2 cards from the 6 cards in which *at least* 1 card is a king.

Therefore, the chance of picking 2 aces is 6 out of 15, which simplifies to 2 out of 5, and the chance of picking 2 cards in which at least 1 card is a king is 9 out of 15, which reduces to 3 out of 5.

Thus, if the mark accepts this bet, they will win on average about 2 times out of 5, which means the con artist will win 3 out of 5 such bets.

The odds are therefore 3 to 2 in favor of the hustler winning the bet.

5

THE GAME OF CHUCK-A-LUCK

"Probability theory is nothing but common sense reduced to calculation."
—*Pierre-Simon Laplace*

PROBLEM*

The game of chuck-a-luck is a famous old swindle. It is usually played at fairs and attractions throughout the United States. The operator has a device that looks like an upturned bird cage. Inside the cage there are 3 large, 6-sided dice. Spectators are urged to get involved in the betting game by the operator and their cronies.

However, sometimes the game is played by enterprising con artists in bars or other public places using just 3 normal, 6-sided dice.

The betting game proceeds as follows: The operator rolls 3 dice. Suppose the top faces on the 3 dice when rolled are 2, 4, and 6.

Those punters in the bar who bet on the number 2 will get even money for their bet. In other words, if the punter bet $1 on number 2, they would win $1 and would also receive back the $1 they waged.

The same applies to the gamblers who bet on the numbers 4 and 6. They would win at odds of even money, plus receive the amount they bet in return. Those punters who bet on the numbers 1, 3, and 5 lose their stakes, of course, and so do not receive any winnings whatever.

If a punter bets any 1 of the 6 numbers and the top faces on 2 of the dice roll to that number, the pundit is paid at odds of 2 to 1. Thus, if they had put $1 on say, number 5, and 2 of the top faces of the 3 dice show a 5, then that bettor wins $2 and receives their stake, which is $1, back in return.

If a punter bets $1 on any 1 of the 6 numbers and the top faces on 3 of the dice roll to that number, the pundit is paid at odds of 3 to 1. Thus, if they had put $1 on say, number 2, and the top faces of the 3 dice show a 2, then that bettor is paid $3 and receives their bet back, which is $1.

The game is appealing to many who like a flutter, because in nearly half of the rolls (91 of 216 rolls) some lucky punter wins something.

The question is this: If a stranger approaches you and offers you an opportunity to play this game, should you accept?

If you do accept, what is the probability of winning at chuck-a-luck?

---

*Source: Martin Gardner, *More Mathematical Puzzles and Diversions* (Harmondsworth, UK: Penguin Books, 1966), 113–117.

SOLUTION

The first of the 3 dice rolled can fall in any 1 of 6 ways. The second dice rolled can fall in any 1 of 6 ways. So, 2 dice can fall in any 1 of 36 different ways. The third dice can also fall in any 1 of 6 ways. Thus, the total number of ways the 3 dice can fall is 6 × 6 × 6, or 216 ways.

The probability that the top faces of the 3 dice do NOT show a similar face is ⅚ × ⅚ × ⅚. This equals ¹²⁵⁄₂₁₆. In other words, there are 125 ways 3 dice can be rolled so that *none* of the top faces of the 3 dice are matching.

Consider the probability of getting only *1* correct number up. The probability of this occurring is ⅙ × ⅚ × ⅚, which equals ²⁵⁄₂₁₆. The incorrect number can be on any 1 of the 3 dice, so we multiply ²⁵⁄₂₁₆ by 3 to get ⁷⁵⁄₂₁₆, which is the probability we seek. Therefore, if a punter bets on any 1 of the 6 numbers available, there are 75 ways out of 216 in which the 3 dice can be rolled so that that selected number will show on 1 of the 3 dice.

Consider the probability of getting *2* correct numbers up. The probability of this happening is ⅙ × ⅙ × ⅚, which equals ⁵⁄₂₁₆. The incorrect number can be on any 1 of the 3 dice, so we multiply ⁵⁄₂₁₆ by 3 to get ¹⁵⁄₂₁₆, which is the probability we seek.

Of course, the probability that the top faces of the 3 dice will all be the same is ⅙ × ⅙ × ⅙, which equals ¹⁄₂₁₆.

We have now obtained the following probabilities in a single game of chuck-a-luck:

Probability of rolling 3 dice so that there are *no* matching numbers	125 chances in 216
Probability of rolling 2 matching numbers	15 chances in 216
Probability of rolling 1 matching number	75 chances in 216
Probability of rolling 3 identical numbers	1 chance in 216
Total	216 chances in 216

Let's assume the punter (that's you!) is playing 216 games of chuck-a-luck, and you are betting $1 at every game. From the above figures you can expect on average to lose 125 games. On these occasions you win nothing and lose your stake.

You can expect on average to roll 1 matching number 75 times in every 216 rolls. On each of these occasions you will be paid $1, plus your stake of $1 back. Therefore, on each of these 75 occasions you will receive $2, or a total of $150.

You can expect to roll 2 matching numbers 15 times in every 216 rolls. On each of these 15 occasions you are paid $2, plus your stake of $1 back. Thus, you receive $45.

When you match 3 numbers of the rolled dice you get $3 plus your stake of $1 back. Therefore, you receive $4. This will happen on average once in the course of 216 games.

You have paid $216 to play the 216 games. You receive on average $150 plus $45 plus $4, or $199 for the privilege of playing 216 games at $1 per game. Thus, for every $1 you bet, your expectation of winning is $199/216$ dollars, which equals $0.921296+$. You can therefore expect to lose $0.078703+$ for every $1 you bet. In other words, for every $1 you bet on a game of chuck-a-luck, you can expect to lose about 7.8 cents.

Your loss is, of course, the swindler's gain. Thus, the operator of the swindle can expect to make a profit of about 7.8 cents for every $1 bet on the game. That is a nice little earner for the hustler.

To illustrate further how lucrative this game can be for the swindler, suppose you are just 1 of 6 players playing chuck-a-luck in the bar. Furthermore, suppose each of the 6 players plays at the rate of 1 game every minute. That equals 6 games per minute. Thus, there are $6 bet every minute or $360 bet every hour. The con artist can expect to win 360 times 7.87 cents every hour. This works out at $28.33 profit every hour for the hustler. That is a nice little profit for the hustler for one hour's work.

Avoid this game if you want to hold on to your hard-earned dollars.

6

TURNING OVER AS MANY $5 BILLS AS POSSIBLE

"Misunderstanding of probability may be the greatest of all impediments to scientific literacy."—*Stephen Jay Gould*

PROBLEM

One day a stranger may gather 4 people around a table and propose the following proposition bet: they will ask each player to put forward 2 $5 bills and place the bills in a circle around a table in front of them. They put forward 2 $5 bills also.

All 10 of the $5 bills are placed so that each has the portrait of Abraham Lincoln facing upward, and the Lincoln Memorial depiction facing downward.

The stranger will now propose the following game. Each player in turn is to do the following: Starting at any $5 bill they are to count 1, 2, 3, 4, pointing to each $5 bill as they count each bill in either a clockwise or counterclockwise direction. When they reach 4, they are to turn over that bill.

They then start at any other $5 bill and count to 4 again. They then turn over the bill they land on at the count of 4. In the counting procedure the player can include a turned-over bill in the count. However, they are not allowed to start the count at a turned-over bill.

The object of the game is to end up with as many $5 bills turned over as possible. The winner collects all $50 on the table.

The stranger will allow each of the 4 players to take a turn. Some of the players may achieve 5 or 6 turned over bills, and then find that they cannot proceed any further.

The stranger will then take his turn. He succeeds in achieving 9 turned-over bills out of the 10 bills on the table and wins the bet. He walks away with a cool $40 in profit.

What strategy does the stranger use to succeed in turning over 9 of the 10 bills?

SOLUTION

The strategy is simple. The stranger starts at any $5 bill, and counts 1, 2, 3, 4 (in a clockwise or counterclockwise direction), going from bill to bill as they count to 4. They turn over the bill reached at the count of 4. They now ensure that whatever bill they started at will be the *last* bill they end on with the second count. They duly turn over that bill. For the third count they ensure that whatever bill they started at for the second count will be the *last* bill they end the third count at. And so on, each time ending their count on the bill that began the previous count. Following this strategy, the stranger will succeed in turning over 9 of the 10 $5 bills.

It is a simple but very effective swindle.

Don't get caught by it!

GUESSING 3 CARDS WRITTEN DOWN

"There is more probability of aliens than there is of god so why is it that when someone says they believe in aliens, they are completely crazy. But if they believe in some all-powerful invisible dude in the sky, they are on the road to success?"—*unknown*

PROBLEM

The proposition is that you secretly write the names of any 3 cards from one entire suit of cards, and the con artist will then—in 3 guesses—attempt to name at least 1 of your 3 chosen cards.

A stranger may approach you at a social gathering and offer you the following proposition bet. They will give you the 13 cards of one suit and ask you to examine them and make sure all 13 cards are there.

When you have done that the con artist will ask you to secretly write the names of any 3 cards from the suit on a sheet of paper. You are the only person who knows the identity of these 3 cards.

The con artist then says that if they are allowed 3 guesses, they will attempt to name at least 1 of the 3 cards you have secretly written. The hustler then says that to make the experiment interesting, they propose to bet $10 at even money odds that they will succeed in naming at least 1 of the 3 cards you selected.

The con artist places a $10 bill on the table and asks you to match it.

You sit there, probably feeling a little bewildered by the con artist and their unusual proposition. You know that there are 13 cards in every suit. You have secretly written the names of 3 of these cards. Only you know the identity of these 3 cards. Yet this stranger is telling you that if they are allowed only 3 guesses, they will name 1 of the 3 cards that you hold. The stranger is even willing to bet $10 that they will succeed! You are thinking that every knowledgeable gambler who has ever held a deck of cards in their hand must know that the odds in this bet must be against this stranger.

You eventually reason that you will accept this little bet and make yourself a handy $10.

Of course, this is precisely the way the con artist wants you to reason.

The question is this: Should you accept the wager?

SOLUTION

No way!

The con artist will win this bet in about 3 out of every 5 wagers they make.

To calculate the true odds of this proposition bet, let us look at the probability that the con artist will *not* correctly guess any of your 3 cards.

In their first guess, the probability that the con artist will *not* correctly guess any of the cards you have written is $^{10}/_{13}$. (Recall that if they are *not* to guess your 3 cards, they have only 10 possibilities to name a card; therefore, their odds of doing this and *not* matching your 3 cards is $^{10}/_{13}$.) The probability that their second guess will *not* correctly match any of your cards is $^{9}/_{12}$, and the probability that their third guess will *not* match any of your cards is $^{8}/_{11}$.

These 3 fractions are then multiplied together to obtain the overall probability that the hustler will NOT guess any 1 of your 3 cards:

$$\frac{10}{13} \times \frac{9}{12} \times \frac{8}{11} = \frac{720}{1716} = 0.4195+$$

We find the probability that the con artist's guesses do NOT match any of your 3 cards is 0.4195+. To find the probability that at least 1 of the con artist's guesses DOES match 1 of your 3 cards we subtract 0.41958+ from 1 (*certainty*).

So $1 - 0.4195+ = 0.5804+$, which in percentage terms equals 58.04 percent.

Thus, the probability that the con artist will—in 3 guesses—name at least 1 of your 3 cards is 58.04 percent.

The odds favor the hustler winning this bet. For every 100 such bets at $10 a go, the con artist can expect to win about 58 of them. Thus, for every $1,000 bet, the scammer can expect to win $580 and lose $420. This leaves them with a profit of $160. That works out as a profit of 16 percent for the grifter.

THE 3 KINGS, 3 QUEENS, AND 3 JACKS PROPOSITION BET

"The excitement that a gambler feels when making a bet is equal to the amount he might win times the probability of winning it."—*Blaise Pascal*

PROBLEM

You may be in a bar one night when a stranger asks you to take 3 kings, 3 queens, and 3 jacks from a deck of 52 cards.

The stranger will then ask you to shuffle these 9 cards and lay them facedown in a row on a table.

The hustler will then ask you to pick any 2 cards. They will bet you $10—at even money odds—that 1 of the 2 cards you pick will be a jack.

If you are contemplating accepting the bet, you will probably reason as follows: there are 9 cards facedown on the table. Six of these cards are not jacks. Surely it is more likely that of the 2 cards I will pick, none of the 2 cards will be a jack.

Of course, this is precisely the way the hustler wants you to reason.

Should you accept the bet?

SOLUTION

No way! The bet favors the con artist.

Look at it this way. There are 9 cards facedown on the table. There are 9C2 or 36 ways of choosing 2 cards from the 9 cards.

If we discard the 3 jacks for a moment, we are left with 6 cards: 3 kings and 3 queens. There are 6C2 or 15 ways of choosing 2 cards from these 6 cards. Thus, there are just 15 ways of selecting 2 cards where none of those 2 cards are jacks.

But as we have seen, there are 36 ways of selecting 2 cards from the 9 cards. Therefore, there are 36 – 15 or 21 ways of picking 2 cards from the 9 cards where *at least* 1 of the cards is a jack.

Therefore, when you choose 2 facedown cards, the odds that at least 1 of them is a jack are 21 to 15. These odds simplify to 7 to 5 in favor that you will choose at least 1 jack.

In other words, the probability you will choose at least 1 jack among the 2 cards you pick is 58.33+ percent. The probability that you will NOT choose a jack when you select 2 cards is therefore 100 – 58.33+ = 41.66+ percent.

The bet massively favors the hustler.

PROPOSITION WITH 2 HALF-DOLLARS

"If you don't know what you want, you will probably never get it."—*Oliver Wendell Holmes Jr.*

PROBLEM

This proposition involves 2 half-dollars, 1 of which is 2-sided.

You may be enjoying a drink in your favorite watering hole when a well-dressed, attractive con artist enters the bar, orders a drink, and gets chatting with other customers. Eventually the subject turns to gambling. The stranger pulls 2 half-dollars from their pocket. They show that 1 of the coins is a normal half-dollar and the other coin is a 2-sided half-dollar (in other words the coin has 2 heads).

The hustler produces a little bag and asks the mark to check if it is empty. The mark confirms that the bag contains nothing. The con artist then drops the 2 half-dollars into the bag and shakes the bag so that the 2 coins are mixed. They then take 1 coin from the bag and toss it in the air. The coin falls with a head facing up.

The attractive con artist then offers the following bet to the mark. They bet $10, giving odds of 5 to 4 in the mark's favor, that the other side of the coin they tossed is heads also.

Should the mark accept the bet?

SOLUTION

It is best not to accept this wager, as the odds favor the hustler.

Consider the genuine coin. One side is heads. Call that side Head1. The other side of the genuine coin is, of course, tails.

Consider the 2-headed coin. Let us imagine 1 side of this coin as being marked Head2 and the other side of the coin as being marked Head3.

We are told that when the hustler takes 1 coin from the bag and tosses it, that the coin falls heads facing up.

There are 3 possibilities concerning this outcome.

The first possibility is that it is the genuine coin. If the side facing up is heads1, the other side of the coin is, of course, tails.

The second possibility is that the 2-headed coin was tossed and the side that is facing up is heads2. The other side of this coin is heads also.

The third possibility is that the 2-headed coin was tossed and the side facing up is heads3. The other side of the coin is heads also.

These are the only 3 possibilities. We see that in 2 of the 3 possibilities the other side of the coin is heads.

Therefore, when the hustler tosses 1 coin and it falls heads, the probability that the other side of the coin is heads also is ⅔.

The bet favors the con artist.

In the long term, the hustler can expect to win this wager 2 times out of every 3 bets. Thus, even if the hustler is offering odds of 5 to 4, they will still make a nice profit if they are regularly performing this bet.

Suppose that over a period, the con artist performs this bet 100 times at $10 a bet. They will win the bet about 66 times. This will bring them a profit of $660 dollars. When they lose the bet they pay out $12.50. In 100 bets they will do this about 33 times. Thus, their losses are $12.50 × 33 = $412.50. That still leaves the con artist with a profit of $247.50, over the course of 100 bets, at $10 per bet.

Thus, regardless of how attractive the hustler may be, do not accept this bet from them!

TURNING OVER 10 CARDS AND BETTING THERE IS AN ACE

"Natural selection is a mechanism for generating an exceedingly high degree of improbability."—*Ronald Aylmer Fisher*

PROBLEM

The proposition involves the turning over of the top 10 cards of a deck and betting that there is an ace among those top 10 cards.*

A con artist may approach you one day at a carnival or some other occasion and offer the following proposition bet to you: they will ask you to shuffle a normal deck of 52 cards, and then lay them facedown on the table in front of you.

The swindler will probably point out that there are 4 aces in a deck. They will probably also mention that since there are 52 cards in a deck, one would need to turn over 13 cards from the top of a shuffled deck to have an even chance of finding an ace.

The hustler will then say they are feeling lucky. Consequently, they propose that they will turn over only the first 10 cards of the deck and bet $10 (at even money odds) that 1 of those cards will be an ace.

Should you accept the bet?

---

*Source: Oswald Jacoby, *How to Figure the Odds* (Garden City, NY: Doubleday, 1947), 114.

SOLUTION

No, do not accept this bet.

The probability of turning up at least 1 ace when turning over the top 10 cards of a shuffled, 52-card deck is 58.65 percent.

The bet therefore favors the con artist.

This surprising probability result is obtained as follows: The deck consists of 52 cards. We remove the 4 aces, leaving 48 cards. We now calculate how many ways 10 cards can be selected from a deck of 48 cards. This is usually written mathematically as 48*C*10. This equals 6,540,715,896. Each of those 6,540,715,896 groups of 10 cards does *not* contain an ace.

The number of different ways 10 cards can be chosen from 52 cards is 52*C*10. This equals 15,820,024,220. Each of those 15,820,024,220 groups of 10 cards *may* contain at least 1 ace.

The difference between 15,820,024,220 and 6,540,715,896 is 9,279,308,324. This is the number of groups of 10 cards that *contain* at least 1 ace.

The probability therefore that the first 10 cards on top of a shuffled deck contain at least 1 ace is 9,279,308,324 divided by 15,820,024,220. This equals 0.586554621+. In percentage terms this equals 58.6554+ percent.

Thus, in percentage terms the probability that the top 10 cards of a shuffled deck contain an ace is 58.6554+ percent. The probability the top 10 cards do *not* contain an ace is 100 − 58.6554 = 41.3446+ percent.

As you can see the odds clearly favor the hustler! For every 100 bets at $10 per bet, the con artist profits to the tune of $587 and loses $413. Thus, in 100 such bets, the hustler makes a profit of $174.

Incidentally, if the hustler cannot get a taker on this bet, they will offer to turn up just 9 cards of a shuffled deck and bet at even money that there is at least 1 ace among the top 9 cards. The probability this time is not as high (in favor of the fraudster) as when 10 cards are turned, but the swindler is still on to be a winner. In percentage terms, the probability that the top 9 cards of a shuffled deck contain at least 1 ace is 54.41 percent.

Even in this version of the proposition bet, the odds still strongly favor the hustler! For every 100 such bets at $10 per bet the con artist wins about $544 and loses about $456. That is still a profit of $88 for the fraudster in 100 such bets.

JACK OR QUEEN IN TOP 6 CARDS IN DECK

"A reasonable probability is the only certainty."—*E. W. Howe*

PROBLEM

The proposition is turning up the top 6 cards of a deck and stating that there will be a queen or a jack among those 6 cards.

One evening at a social function you may get chatting with an attractive person. Having chatted about various subjects they may bring the conversation around to their love of numbers and gambling.

The con artist will ask you to shuffle a normal deck of 52 cards. When you are shuffling, they will point out to you that the deck contains 52 cards. They will helpfully point out that just 4 of those cards are queens and just 4 of the cards are jacks; thus, the queens and the jacks make up a total of 8 cards in a deck of 52.

Yet, in a spirit of sportsmanship and generosity, they state that they are prepared to offer the following bet to you at odds of even money. They propose that you shuffle a deck of 52 cards and that you then turn up only the first 6 cards on top of the deck. They bet $10 that there will be a queen or a jack among those top 6 cards.

You will probably contemplate the bet for a minute or two. You may well reason as follows: you are being asked to turn up only the top 6 cards of a 52-card deck that has been thoroughly shuffled. Surely, it is unlikely that at least 1 of those 6 cards will be a queen or a jack.

That is the way of reasoning, of course, that the swindler is hoping you have adopted.

Should you accept the bet?

SOLUTION

No way!

The odds of winning the bet are nearly 2 to 1 in favor of the con artist!

Let's calculate how many ways there are of choosing groups of 6 cards from a deck that has the 4 queens and 4 jacks removed from it. In other words, we want to calculate how many ways there are of choosing 6 cards from a deck of 44 cards. This is written mathematically as 44C6. This equals 7,059,052. Call this Set A.

Now we need to calculate how many ways there are of choosing 6 cards from 52. We can write this mathematically as 52C6. This equals 20,358,520. Call this Set B.

If the 6 cards turned up on top of the deck belong to Set A, we know that there is *no* queen or jack among those 6 cards. The probability of this happening is Set A divided by Set B.

In other words, the probability that there is *no* queen or jack among the top 6 cards of the deck is therefore $^{7,059,052}/_{20,358,520}$. This equals 0.3467+.

To find the probability that there is a queen or jack among the top 6 cards of the deck we subtract 0.3467+ from 1 (*certainty*). The answer is 0.6532. In percentage terms, this equals 65.32 percent.

Thus, the probability that a queen or a jack (or any 2 named cards) is among the top 6 cards of a deck is 65.32+ percent.

As you can see the bet favors the hustler. The odds are nearly 2 to 1 in favor of the con artist!

It is wise not to accept this bet if it is offered to you.

12

AN ACE OR FACE CARD ON TOP OF 1 OUT OF 3 STACKS

"If we have an atom that is in an excited state and so is going to emit a photon, we cannot say when it will emit the photon. It has a certain amplitude to emit the photon at any time, and we can predict only a probability for emission; we cannot predict the future exactly."—*Richard P. Feynman*

PROBLEM

The proposition is that there will be an ace, king, queen, or jack on top of at least 1 of 3 piles cut from a shuffled deck.

You may be in a bar having a quiet drink one evening when you are approached by a smooth-talking stranger. The stranger may speak about various topics for a few minutes, but they will then swing the subject of conversation around to the theme of gambling. They will tell you how strange probability theory is and, as an example of just how strange it can be, will ask you if you want a little bet that might put $10 your way.

Intrigued, you will probably ask for more information.

Having captured your interest, the con artist will now attempt to capture some of your hard-earned dollars. The swindler will point out that there are 4 aces, 4 kings, 4 queens, and 4 jacks in a deck of cards. That is a total of just 16 cards out of a deck that contains 52 cards.

Nevertheless, the hustler will say that they are prepared to bet $10 at even money odds that if you shuffle a deck of 52 cards and then cut the deck into 3 piles, the top card of at least 1 of those 3 piles will be an ace, king, queen, or jack.

The bet appears to favor you. You know that there are 52 cards in a deck. Yet this stranger is telling you that if you cut the deck into 3 piles at least 1 of the top cards in those 3 piles will be an ace, king, queen, or jack. That cannot be right, can it? You believe that the wager must favor you. How can it be any other way? It is common sense, after all.

Or is it?

Should you accept the stranger's bet?

SOLUTION

No way.

If you remove the aces and all the face cards from a deck you will have 36 cards left.

There are 36C3 or 7,140 ways 3 cards can be selected from these 36 cards. Call this Set A. All the 3-card groups in Set A do *not* contain any aces or any face cards.

There are 52C3 or 22,100 ways 3 cards can be chosen from 52 cards. Call this Set B. All the 3-card groups in Set B *may* contain any of the aces or any of the face cards.

When the deck is cut into 3 piles and the top card in each pile is turned, those 3 top cards belong to either Set A or Set B above.

The probability those 3 cards belong to a set is $7,140/22,100$. This equals 0.3230.

In other words, the probability there is *not* an ace, jack, queen, or king among the top 3 cards is 0.3230.

Consequently, the probability that there *is* either an ace, jack, queen, or king among those top 3 cards is 0.3230 subtracted from 1 (*certainty*). This equals 0.6770. In percentage terms this equals 67.70 percent.

In other words, the probability that the 3 cards on top of the 3 piles contain either an ace, a king, a queen, or a jack is 67.70 percent.

The probability that the 3 cards on top of the 3 piles do *not* contain an ace, a king, a queen, or a jack is 100 – 67.70 percent. This equals 32.30+ percent.

Therefore, the odds are more than 2 to 1 in the con artist's favor.

This proposition bet is another moneymaker for the hustler!

PROPOSITION INVOLVING 10 CARDS OF 1 SUIT

"I don't believe in providence and fate; as a technologist I am used to reckoning with the formulae of probability."—*Max Frisch*

PROBLEM

A stranger at a gathering one evening may offer you the following proposition.

They will take the following 10 cards of any 1 suit: ace, 2, 3, 4, 5, 6, 7, 8, 9, and 10, and hand them to you.

They will then ask you to thoroughly shuffle the packet of 10 cards and to place them facedown on a table. No one now knows how the 10 cards are distributed.

The swindler will then offer you the following proposition bet: They will ask you to turn over any 6 cards, 1 card at a time. However, you must not turn over the ace. If you do, you lose. If you avoid picking the ace, you win. It's as simple as that!

The hustler will bet you $10 at even money odds that 1 of the 6 cards you turn over will be the ace.

The bet appears tempting. Your line of reasoning may be as follows: There are 10 cards facedown on the table. Only 1 of them is the ace. I am being asked to choose any 6 of these 10 cards. Thus, it must be more likely than not that I will avoid the ace. The probability must be in my favor that I will avoid the ace.

Of course, this is the way the con artist wants you to reason.

The question is this: Should you accept this wager?

SOLUTION

No way!

The probability of picking 6 cards, *none* of which is the ace, is calculated as follows:

$$\frac{9}{10} \times \frac{8}{9} \times \frac{7}{8} \times \frac{6}{7} \times \frac{5}{6} \times \frac{4}{5} = \frac{4}{10}$$

All the numerators and denominators cancel except the 4 and 10. Therefore the probability that you will *not* pick the ace is 4/10. Consequently, the probability you *will* pick the ace is 6/10. This equals 60 percent.

The odds favor the hustler 60 to 40, or 3 to 2, in this wager. In other words the hustler will win this bet 3 times out of every 5 wagers.

If you accept this bet at even money odds, the con artist will laugh all the way to the bank.

PROPOSITION INVOLVING 2 FAIR, 6-SIDED DICE

"It is scientific only to say what's more likely or less likely, and not to be proving all the time what's possible or impossible."—*Richard P. Feynman*

PROBLEM

You may be watching a game of craps one day when a stranger approaches you and offers you the following bet: They say you can throw a pair of dice that you choose, but before you do they will name 2 numbers from 1 through 6. They will bet you $5 at even money odds that at least 1 of those 2 numbers will be the top-most number on 1 of the 2 dice rolled.

Your line of thinking will probably lead you to the conclusion that it is unlikely that at least 1 of the 2 numbers named by the stranger will be rolled, given the fact that each 6-sided die is fair. Therefore, you will most likely be inclined to believe that the odds in this little wager favor you. Consequently, you may be inclined to accept the bet.

Should you accept this proposition bet?

SOLUTION

This is a classic betting proposition. The odds favor the hustler. In fact, the odds are 5 to 4 in favor of at least 1 of the 2 named numbers being rolled.

The odds that 1 of 2 named numbers—on 1 die—will *not* be rolled when 1 die is thrown is $\frac{4}{6}$. Consequently, when 2 dice are thrown, the odds that 2 named numbers on each of the 2 dice will not be thrown is $\frac{4}{6} \times \frac{4}{6}$, which equals $\frac{16}{36}$.

To obtain the probability that at least 1 of 2 numbers *will* be rolled we subtract $\frac{16}{36}$ from 1 (*certainty*). This equals $\frac{20}{36}$.

Therefore, the probability is $\frac{20}{36}$ that *at least* 1 of the 2 numbers named by the hustler *will* be rolled. This probability simplifies to $\frac{5}{9}$.

Thus, there are 5 chances in 9 that at least 1 of the 2 named numbers will be rolled. Consequently, there are 4 chances in 9 that at least 1 of the 2 named numbers will *not* be rolled.

The odds therefore are 5 to 4 in favor of the con artist winning the bet.

In percentage terms the hustler will win this bet 55.55 percent of the time.

It is sound advice to steer clear of this wager if it is offered to you.

MATCHING CARDS FROM 2 SIMULTANEOUSLY DEALT DECKS

"The probability that we may fail in the struggle ought not to deter us from the support of a cause we believe to be just."—*Abraham Lincoln*

PROBLEM*

The proposition is that when 2 shuffled decks of cards are dealt simultaneously at the rate of 1 card at a time, at least 2 cards, 1 from each deck, will match.

One evening a stranger may take 2 decks of cards from their pockets and ask you to examine and then shuffle each of the 2 decks. They will then ask you for 1 of the decks.

The con artist now proposes the following bet: They state that you both deal 1 card at a time from the top of each respective deck, checking for a perfect match (both suit and rank). They state that you will both perform this routine on 3 occasions. They then bet you $10 at even money odds that at some point over the 3 games there will be a perfect match.

Should you accept the bet?

---

*Source: Martin Gardner, *Further Mathematical Diversions* (Harmondsworth, UK: Pelican Books, 1977), 40.

SOLUTION

No way!

The mathematics used in determining the probability of this wager is beyond the scope of this book. But it can be worked out that the probability of getting a match is 63.2120+ percent. In other words, in the long run, in nearly 2 out of every 3 bets the procedure will result in a match.

Beginning with 1 card, 2 cards, 3 cards, and so on, the number of ways of NOT getting a match is as follows:

NUMBER OF CARDS	NUMBER OF WAYS TO NOT GET A MATCH
1	0 out of 1 possible arrangements
2	1 out of 2 possible arrangements
3	2 out of 6 possible arrangements
4	9 out of 24 possible arrangements
5	44 out of 120 possible arrangements
6	265 out of 720 possible arrangements
7	1,854 out of 5,040 possible arrangements
8	14,833 out of 40,320 possible arrangements
9	133,496 out of 362,880 possible arrangements

The above numbers, 0, 1, 2, 9, 44, and so on, are obtained as follows: Multiply each successive number in the series by 2, 3, 4, 5, and so on. When the multiplier is even, add 1 to the result. When the multiplier is odd, subtract 1 from the result. The answer will be the next successive number in the series.

Thus, your calculations are $2 \times 0 + 1 = 1$; $3 \times 1 - 1 = 2$; $4 \times 2 + 1 = 9$; $5 \times 9 - 1 = 44$. And so on.

The numbers 1, 2, 6, 24, 120 on the far right-hand side in the table above are the factorial numbers. In mathematics, 0 factorial = 1; 1 factorial = 1; 2 factorial = 2; 3 factorial = 6; 4 factorial = 24; 5 factorial = 120, and so on. (The n factorial equals n multiplied by all the integers equal to or less than n. For example, 3 factorial equals $3 \times 2 \times 1 = 6$. Similarly, we find that 4 factorial equals $4 \times 3 \times 2 \times 1 = 24$.)

The probability of *not* getting a match with 1 card is 0, with 2 cards is ½, with 3 cards is ⅔, with 4 cards is $9/24$, with 5 cards is $44/120$, with 6 cards is $265/720$, with 7 cards is $1854/5040$, and so on. The value of each successive fraction is approaching the value of 0.367879441.

Consequently, to find the probability that there will be a match we subtract 0.367879441 from 1 (*certainty*) and obtain 0.632120558.

Therefore, the probability of obtaining a match with 1 card is 0, with 2 cards is ½, with 3 cards is ⅚, with 4 cards is $15/24$, with 5 cards is $76/120$, with 6 cards is $455/720$, with 7 cards is $3186/5040$, and so on. The value of each successive fraction is approaching the value of 0.632120558.

In other words, as the number of cards increase toward infinity the probability that there will be a match approaches (but never exceeds) 0.632120558. This number equals $1 - (1/2.718281828)$. That number in the parenthesis, 2.718281828, is a famous number, well known to mathematicians. It is usually denoted as e. It crops up in many areas of mathematics.

Consequently, we find that in a deck of 52 cards the probability of a match is 63.2120+ percent. This nearly equals ⅔. Thus, in the long term, the hustler will win this bet 2 out of every 3 times they offer the bet. Therefore, if you play 3 of these games with the con artist, the odds are in their favor that they will win 2 of them.

Incidentally, one astonishing fact can be gleaned from this proposition bet. As the number of cards in the deck increases, the possibility of achieving any match decreases *if no other factors are considered*. However, as the number of cards increases the number of possible locations for matching cards within the deck also increases. It is a remarkable fact that as the number of cards increases toward infinity these two factors cancel each other, so that the probability of a match remains at 63.2120+ percent.

As with many other gambling propositions, the odds in this wager favor the hustler.

16

CLASSIC PROPOSITION WITH 5 RED CARDS AND 2 BLACK CARDS

"If we look at the way the universe behaves, quantum mechanics gives us fundamental, unavoidable indeterminacy, so that alternative histories of the universe can be assigned probability."—*Murray Gell-Mann*

PROBLEM

One evening a stranger may approach you in a bar and offer you the following proposition bet: They will show you 5 red cards and 2 black cards and ask you to shuffle the 7 cards. They will then ask you to lay each of the cards facedown in a row. At this point neither of you know which of the 7 cards are red or black.*

The hustler may then say something along the following lines to you: "We have here 7 cards. Five of them are red and only 2 of them are black. I am going to ask you to choose 3 cards, 1 at a time. If all 3 cards are red, I will give you 20 bucks. However, in the unlikely case you happen to choose 1—or 2—cards that are black, then you, my friend, will give me 20 bucks. That is fair, is it not?"

The hustler will then take $20 from their wallet and lay it down on a table near you. They will ask you to cover the bet—that is, they will ask you to place 20 of your dollars on top of theirs. The winner of the bet will then take the $40 on the table.

The swindler may also say something along the following lines: "This bet is quite attractive to you, my friend. You see, when you make your first choice, the odds are 5 to 2 in your favor that you will choose a red card. Suppose you pick a red card. In that case when you make your second choice the odds are 4 to 2 in your favor, which work out at 2 to 1, that you will pick a red card. If you choose a red card this second time the odds are still in your favor when you make your third choice, 3 to 2 in your favor, in fact, that you will choose a red card. Thus, the odds favor you at every stage of the game. I ask you: What can be fairer than that? Now here is the thing. I wish to do a little experiment. I am going to offer you odds of even money that you will *not* pick 3 red cards. Yes, odds of even money! I must be mad! Where would you get it? You will not get a better deal than this on this side of the Atlantic Ocean, my friend! I say 20 bucks you will not choose 3 red cards. That's the bet! So, what are you waiting for? I have put my $20 down. You will never again be offered an attractive deal like this. Grab the opportunity when you can!"

You are sitting there wondering what to do next. The con artist is very persuasive, and you're just wondering if the bet favors you.

Should you agree to accept this wager?

---

*Source: Nick Trost, *Expert Gambling Tricks* (Columbus, OH: Trik-Kard Specialties, 1975), 19–20.

SOLUTION

You would be very wise to avoid this betting game!

The hustler's calculation of the odds is very plausible and has fooled many a mark over the years. The con artist's calculations are correct *if* you choose a red card at every stage of the game.

To calculate the probability that you will choose 3 red cards we first calculate how many ways 3 cards can be chosen from 7. This is mathematically written as $7C3$. This equals 35.

Now we need to calculate how many ways 3 cards can be chosen from 5. This is mathematically written as $5C3$. This equals 10.

Now we can calculate the true odds of this famous swindle.

When the 7 cards are placed facedown on the table, we are asked to select 3 cards—1 at a time—from the 7. This can be done in 35 ways. Some of these 3-card selections will include only 3 *red* cards. But some of these 3-card selections will also include 1, or 2, red cards, plus a black card, or in some cases, 2 black cards.

On the other hand, when we select 3 cards from the 5 red cards, *all* the 3-card selections will consist of red cards only. We find that there are only 10 of these selections possible.

To obtain the number of 3-card selections that contain at least 1 *black* card we subtract 10 from 35. The answer is 25.

Thus, we have established the following facts: There are 25 3-card selections that contain at least 1 black card. Call this set of selections Set A. There are 10 3-card selections that do not contain a black card. Call this set of selections Set B.

When the hustler offers you the opportunity to select 3 cards from the 7 cards facedown on the table, it is much more likely that the 3 cards you select will belong to Set A.

In fact, the odds that the 3 cards you select will belong to Set A are 25 to 10, which simplify to 5 to 2.

Consequently, you have only 2 chances in 7 of winning this bet. The hustler has 5 chances in 7 of winning the proposition. The odds clearly favor the hustler.

For every 7 bets at $20 per bet, the hustler wins $100. On the 2 occasions they lose, they lose $40. Thus for every $140 bet by the fraudster, they gain $100. That means the mark loses $100 and gains $40.

The bet clearly favors the con artist.

Avoid this bet if you want to hold on to your hard-earned dollars!

17

ROLLING ONE 6 IN 3 CONSECUTIVE DIE ROLLS

"As a doctor, as a man of science, I can tell you there is no such thing as curses. Everything just happens as a question of probability. The statistical likelihood of a specific event."—*Andrew Schneider*

PROBLEM

You may be at a carnival or midway one day and a stranger may ask you if you want a little bet involving just one 6-sided die. They state that they will even let you throw the die several times to let you see that the die is a fair one.

Before offering the bet to you they helpfully point out that there are 6 numbered faces on the die and that 1 of the 6 faces is the number 6. They then state that according to the laws of probability you have 1 chance of rolling a 6 when you throw 1 die once. But they will then inform you that when you throw 1 die 3 consecutive times your chances of throwing a 6 are now equal to $\frac{1}{6} + \frac{1}{6} + \frac{1}{6}$, or $\frac{3}{6}$. They will then say that when they went to school, they were taught that $\frac{3}{6}$ is equal to $\frac{1}{2}$. Therefore, when you roll 1 die 3 times, the laws of probability state that there is 1 chance in 2 that you will roll a 6 on at least 1 of those 3 rolls.

The con artist then bets you $5 at odds of 5 to 4 in your favor that you will *not* roll exactly one 6 on any 1 of the 3 rolls.

You are tempted to take the bet, particularly as the odds the con artist is offering seem to favor you.

Should you accept the bet?

SOLUTION

If you accept this bet at these odds you are being conned by the fraudster. The odds in this classic proposition bet are in favor of the hustler.

Here is the correct way to calculate the probability of throwing *exactly* one 6 on three rolls of 1 die: The chance that a 6 will be rolled on 1 roll is $\frac{1}{6}$.

This must then be multiplied by $\frac{5}{6}$ and then by $\frac{5}{6}$ again (the chance that one 6 will *not* be rolled on either of the two other rolls) and this then must be multiplied by 3 (because any 1 of the 3 rolls of the die could produce a 6).

The entire calculation is: $\frac{1}{6} \times \frac{5}{6} \times \frac{5}{6} \times 3$, which equals 0.3472+.

This is the probability of rolling *exactly* one 6 with 3 rolls of 1 die. If we subtract 0.3472+ from 1 (*certainty*) we obtain 0.652777+, which is the probability that one 6 exactly will *not* be rolled when rolling 1 die 3 times. This probability in percentage terms equals 65.2777+ percent.

Thus, the odds favor the hustler. In the long run, they can expect to win this little wager about 2 times out of every 3 bets.

The con artist can thus afford to give odds of 5 to 4 with this wager.

Let us say they perform 100 of these wagers at $5 a bet. On every 100 bets they will win about 65 percent of the time, giving them 65 × $5 or $325. They will lose about 35 percent of the time. Each time they lose they drop $\frac{5}{4}$ dollars, which equals 6.25 greenbacks. With 35 losses they drop 218.75 bucks. That still leaves the hustler with a tidy profit of $106.25. The con artist may even make more than this if they manage to get suckers to take the bet at even money odds. Their profit then would be $150 for 100 bets at $5 a go.

In any language, that is a tidy profit for the hustler.

18

DEALING 6 CARDS THAT INCLUDE A PAIR

"Million-to-one odds happen eight times a day in New York."—*Penn Jillette*

PROBLEM*

A friendly looking person will walk into a bar some evening when you are enjoying a social drink and engage you in conversation.

Eventually the friendly stranger will bring the conversation around to gambling. At some stage of the evening the stranger (or con artist) will politely ask you to shuffle a 52-card deck. They will conveniently point out to you that it has been mathematically proven that there is less than an even chance that a pair will be dealt if one deals 5 cards from a shuffled deck. (A pair is a hand that contains 2 cards of 1 rank and 3 cards of 3 other ranks. The probability of obtaining a pair when 5 cards are dealt from a 52-card deck is 42.2569+ percent.) The stranger will then go on to offer you the following bet: deal just 6 cards from a shuffled deck and the con artist bets you $10 at 6-to-5 odds in your favor that among those 6 cards a pair will be dealt. In other words, if you win the bet you will collect $12 and get your $10 back.

You contemplate the bet. You reason that if the odds are less than even that a pair will appear when 5 cards are dealt, they must be close to even if 6 cards are dealt. The stranger is offering odds of 6 to 5 that a pair will show. Consequently, your reasoning will tell you that the bet favors you.

Should you accept the bet?

---

*Source: Oswald Jacoby, *How to Figure the Odds* (Garden City, NY: Doubleday, 1947), 115.

SOLUTION

No way! The odds in this bet are strongly in favor of the hustler.

There are 52 ways the hustler can deal the first card, 51 ways of dealing the second card, 50 ways of dealing the third card, 49 ways of dealing the fourth card, 48 ways of dealing the fifth card, and 47 ways of dealing the sixth card. Thus, the total number of ways of dealing 6 cards from a 52-card deck is $52 \times 51 \times 50 \times 49 \times 48 \times 47$. This answer must then be divided by 6 factorial, which equals 720, because the *order* of the cards is not considered. (The expression we call 6 factorial is usually written as 6! It equals $6 \times 5 \times 4 \times 3 \times 2 \times 1$, or 720.)

We find that the total number of ways of dealing 6 cards from a 52-card deck is 20,358,520.

Now we calculate the number of ways of dealing 6 cards so that there is *no* pair. The first card dealt can be any 1 of 52 cards. The second card dealt must be 1 of the other 48 cards if *no* pair is to be dealt. Similarly, the third card dealt must be 1 of the other 44 cards if *no* pair is to be dealt. And so on. The entire calculation is $((52 \times 48 \times 44 \times 40 \times 36 \times 32)/720)$, which equals 5,060,689,920/720 or 7,028736.

Thus, our calculation tells us that there are 7,028,736 ways of dealing 6 cards from a 52-card deck so that there is *no* pair among the 6 cards.

Therefore, the number of 6-card hands that contain a pair must equal 20,358,520 − 7,028,736. This equals 13,329,784.

Thus, when 6 cards are dealt from a 52-card deck, the probability that a pair is among those 6 cards is $^{13,329,784}/_{20,358,520}$. This fraction equals 0.65475211. Thus, the probability of getting a pair is 65.4752+ percent. The probability of *not* getting a pair is 100 − 65.4752+ percent. This equals 34.5248+ percent.

We see that the odds are nearly 2 to 1 in favor of the hustler!

Consequently, the hustler can expect in the long term to win this bet about 2 times out of every 3 wagers offered.

It is a nice little earner for the hustler.

19

DEALING 13 CARDS WITH AT LEAST 1 CARD HIGHER THAN A 9

"Man is descended from a hairy, tailed quadruped, probably arboreal in its habits."
—*Charles Darwin*

PROBLEM

One evening in a bar, a stranger may tell you that they are very interested in betting and in performing experiments connected to gambling. They may add that they would like to try a little experiment with you that involves a normal deck of 52 cards.*

Once they have your interest hooked, they take a deck from their jeans pocket. The con artist will ask you to examine the cards to confirm that it is a normal deck. They will then ask you to thoroughly shuffle the cards.

The con artist will then ask you to deal the top 13 cards facedown from the shuffled deck.

You do as you are asked.

The con artist now offers you the following bet: They say that they bet you $10, giving you odds of 1,000 to 1, that *at least 1* of the 13 facedown cards that you have dealt is a card that is higher than a 9. In other words, *at least 1* of the cards you have just dealt is a 10, jack, queen, king, or ace.

At this stage you are probably sitting there in awe with your mouth wide open and are astounded that this artist is giving odds of 1,000 to 1 on this bet.

The artist may add the following few lines to lure you into accepting the bet. There are 20 cards in the deck that are 10s, jacks, queens, kings, or aces. That means, of course, that there are 32 cards in the deck that are *not* 10s, jacks, queens, kings, or aces. Thus, the friendly con artist will point out that that means the odds favor *no* 10s, jacks, queens, kings, or aces appearing among the 13 cards by 32 to 20, which simplify to 8 to 5. Consequently, they will helpfully tell you there are 5 chances in 13 that you will win.

They point out that this could be your lucky day as they are betting $10 and are offering the astronomical odds of 1,000 to 1 in your favor. If you happen to win the bet, you will win $10,000 and receive your stake of $10 back. If you lose the bet, you have lost just $10. They will insist that you would be crazy not to bet on this proposition.

You are inclined to think you should agree to this bet on the off chance that you just might pick up a handy $10,000 in profit.

The question is this: Should you accept this bet?

---

*Source: Brian Everitt, *Chance Rules: An Informal Guide to Probability Risk and Statistics* (New York: Springer, 1999), 70.

SOLUTION

No way!

This is a famous bet known to bridge players around the world. A hand of 13 cards where no card is higher than a 9 is known as a *yarborough*, because in the nineteenth century the second Earl of Yarborough, Charles Anderson Worsley Anderson-Pelham (1809–1862), was said to have offered the bet to many of his contemporaries (see Alan Truscott, "Bridge; Betting with the Odds," *New York Times*, January 18, 1987). Incidentally, there is no record that the Earl of Yarborough ever had to pay out on the bet.

Here is how we calculate the probability of dealing a *yarborough* from a deck of 52 cards. Since none of the 10s, jacks, queens, kings, or aces can appear in a *yarborough*, we find that we are dealing with a deck of 32 cards. The number of ways 13 cards can be chosen from 32 is 32*C*13. In other words, there are 32*C*13 ways of dealing a *yarborough*. This means that there are 347,373,600 ways a *yarborough* can be dealt.

Consider now the number of different ways of dealing 13 cards from a deck of 52. This can be done in 52*C*13 ways. This equals 635,013,559,600. That is more than 635 thousand million different ways a hand of 13 cards can be dealt from a deck of 52 cards!

Consequently, the probability that a *yarborough* will be dealt is equal to the following fraction: $347,373,600/635,013,559,600$.

This fraction simplifies to 0.0005470333, which is very close to $1/1,828$.

Thus, the probability of being dealt a *yarborough* is about 1 chance in 1,828. Consequently, the odds of being dealt a *yarborough* are about 1,827 to 1 against.

In other words, over the long term, if one deals 13 cards from a shuffled deck one can expect that a *yarborough* will be dealt once in about 1,828 deals.

Therefore, in our example the con artist was very shrewd in offering odds of 1,000 to 1. They offered these apparently generous odds to lure you into accepting the betting proposition.

However, as we can see from the above, in offering such odds, the con artist was shortchanging you. They should have been giving you odds of 1,827 to 1.

If the con artist was regularly performing this proposition bet, at say $10 a go, they could expect in every 1,828 such bets to win about 1,827 of them, netting a tidy $18,270.

On the one occasion that they lose the bet the con artist pays out $10,000. That still leaves the con artist with a handsome profit of $8,270!

In anyone's language, that is nice work, if you can get it!

20

WRITING DOWN 5 LETTERS WITH AT LEAST 1 BEING IDENTIFIED

"My district has been hit with three 500-year floods in the last several years, so either you believe that we had a one in over 100-million probability that occurred, or you believe, as I do, that there's a new normal, and we have changing weather patterns, and we have climate change. This is the science."—*Chris Gibson*

PROBLEM

A stranger may approach you one evening and engage you in conversation about the beauties and intricacies of the English language. They may well steer the conversation in such a way that, before long, you will both be marveling at the wonderful fact that human beings can construct literally hundreds of thousands of different words using just 26 letters of the English alphabet.*

Having manipulated you into this feeling of awe, the stranger may then remark rather casually that they are prepared to make a bet about the English language. The bet is this: They will ask you to secretly write down any 5 letters of the English alphabet. They will then bet you $20—at even money odds—that if they are given 5 guesses, they will be able to name at least 1 of the letters you have written down.

If you think about the bet, you may be inclined to reason as follows: There are 26 letters in the English alphabet. This stranger has asked me to secretly write down any 5 of those 26 letters and bets me that in 5 guesses they will name at least one of the letters I have written down. That surely is unlikely. The odds of this happening must be 5 in 26 chances or about 1 chance in 5. Therefore, I think I will accept the bet, as I think there is a good chance I will win the bet and therefore pick up a handy $20.

That is probably the way you will reason. Of course, the con artist wants you to reason in that way. They, after all, want you to accept the wager.

In any event the hustler places $20 on a nearby table. They ask you to match it. You consider placing your $20 over their $20. The winner of the bet will, of course, take all 40 greenbacks.

The question is this: Should you accept the bet?

---

*Source: Charlie Rice, *Challenge! Fun with Puzzles, Riddles, Word Games and Problems* (Kansas City, MO: Hallmark Cards, 1968), 32.

SOLUTION

This is a beautiful proposition bet. Although the con artist has no way of knowing what 5 letters you have secretly written down, the odds in this bet still favor the hustler.

Let's look at the bet a little more closely. First, we note that at the request of the hustler, you have secretly written down any 5 letters of the English alphabet. Only you know what these letters are.

Let us first calculate the probability that the con artist's 5 guesses will *not* match any of the letters you have chosen.

The con artist chooses 1 letter. It does not matter what letter they choose as it will not affect the odds of this bet. The probability that the hustler's choice is *not* one of your chosen letters is $^{21}/_{26}$. (Remember, you have already accounted for 5 of the letters, so if the con artist is to avoid a match, they cannot name 1 of the letters you have already written down. Therefore, they have 21 chances out of 26 of *not* getting 1 of your chosen letters.) On their second choice, the probability they do not choose 1 of your chosen letters is $^{20}/_{25}$. On their third choice, the probability they do not choose 1 of your chosen letters is $^{19}/_{24}$. On their fourth choice, the probability they do not choose 1 of your chosen letters is $^{18}/_{23}$. On their fifth and final choice, the probability they do not choose 1 of your chosen letters is $^{17}/_{22}$.

All 5 of these fractions must be multiplied together to obtain the probability that the hustler does *not* choose any of your chosen letters when they make their 5 guesses. Thus, we have:

$$\frac{21}{26} \times \frac{20}{25} \times \frac{19}{24} \times \frac{18}{23} \times \frac{17}{22} = \frac{2441880}{7893600} = 0.309349+$$

As we can see this product equals 0.309349+. This is the probability that the fraudster will *not* correctly guess 1 of the 5 letters. We subtract this result from 1 (*certainty*) and obtain 0.690651+. This is the probability that the con artist *will* correctly guess 1 of the 5 letters that you have written down.

In percentage terms the probability of the con artist getting a match is 69.0651+ percent and of *not* getting a match is 30.9349+ percent.

Therefore, the odds are very nearly 7 to 3 that the hustler will get a match.

Thus, in the long run the hustler can expect to win this wager about 7 times in every 10 bets.

This proposition bet is a nice little earner for the con artist.

Don't make life easy for them by accepting their bet.

21

PROPOSITION THAT 1 OF 7 GUESSES (U.S. STATES) WILL BE CORRECT

"If you think you can or you think you can't, you're probably right."—*Henry Ford*

PROBLEM

A stranger may approach you one evening in a bar and engage you in conversation about the joys of gambling.

They may say that they are prepared to make an extraordinary bet.* They reach into one of their pockets and produce a card. On the card are printed the 50 states of the United States.

For the reader's convenience, here are the 50 states of the United States in alphabetical order: Alabama, Alaska, Arizona, Arkansas, California, Colorado, Connecticut, Delaware, Florida, Georgia, Hawaii, Idaho, Illinois, Indiana, Iowa, Kansas, Kentucky, Louisiana, Maine, Maryland, Massachusetts, Michigan, Minnesota, Mississippi, Missouri, Montana, Nebraska, Nevada, New Hampshire, New Jersey, New Mexico, New York, North Carolina, North Dakota, Ohio, Oklahoma, Oregon, Pennsylvania, Rhode Island, South Carolina, South Dakota, Tennessee, Texas, Utah, Vermont, Virginia, Washington, West Virginia, Wisconsin, and Wyoming.

The con artist will ask you to study the card and then to secretly write down any 7 states of the United States on a separate page, and not to let anyone else see what you have written.

The hustler then says that if they are allowed to make 7 guesses, they are prepared to bet $20—at even money odds—that they will correctly guess at least 1 of the states that you have secretly written.

The hustler places $20 on a nearby table. They ask you to "cover the bet." That is a term used by gamblers to ask you to place your stake (in this case, $20) on the table also. The winner of the bet will, of course, collect all $40.

You may well be tempted to reason as follows: This stranger has asked me to secretly write down the names of any 7 states of the United States. I have done that. There is no way they or anyone else can know what states I have listed. Only I know. Yet they say they are prepared to bet $20 at even money odds that if they are allowed 7 guesses, they will name at least 1 of the states I have listed. That seems very, very unlikely, because there are 50 *different* states and only 7 guesses. The odds must be in my favor. I think I will take the bet and if lady luck smiles favorably on me (as I think it will!) I will collect all $40 on the table.

Of course, the con artist is secretly hoping that you will reason like this, because they know that such reasoning will probably lure you into accepting the bet.

If they win the wager, they will make a tidy profit of $20.

The question is this: Should you accept the bet?

---

*Source: Charlie Rice, *Challenge! Fun with Puzzles, Riddles, Word Games and Problems* (Kansas City, MO: Hallmark Cards, 1968), 32. (A slightly different bet.)

SOLUTION

No, you should not accept this bet, as the odds strongly favor the hustler.

At the beginning of the bet you have chosen 7 states of the United States and have secretly written them down. Only you know what 7 states you have selected. But the beautiful thing about this proposition bet is that the con artist need not know what states you have chosen.

Let's analyze the wager a little more closely.

You have secretly written the names of 7 states. The hustler now takes their 7 guesses.

Let us first calculate the probability that their guesses will *not* match any of the states you have chosen. The probability that their first guess will *not* match any 1 of the states you selected is $^{43}\!/_{50}$. (Remember, you have already chosen 7 states, so they cannot choose 1 of those if their choice is *not* to match 1 of the states you have selected.) On their second guess, the probability that their choice will not match 1 of your chosen states is $^{42}\!/_{49}$. On their third guess, the probability that their choice will not match 1 of your selected states is $^{41}\!/_{48}$. On their fourth guess, the probability that their choice will not match 1 of your chosen states is $^{40}\!/_{47}$. On their fifth guess, the probability that their choice will not match 1 of your selected states is $^{39}\!/_{46}$. On their sixth guess, the probability that their choice will not match 1 of your chosen states is $^{38}\!/_{45}$. On their seventh and final guess, the probability that their choice will not match 1 of your chosen states is $^{37}\!/_{44}$.

All seven of these fractions must be multiplied together to obtain the probability that any 1 of the hustler's 7 guesses will *not* match any 1 of the 7 states you have chosen.

$$\frac{43}{50} \times \frac{42}{49} \times \frac{41}{48} \times \frac{40}{47} \times \frac{39}{46} \times \frac{38}{45} \times \frac{37}{44} = \frac{162,409,534,560}{503,417,376,000} = 0.322614$$

Thus, the probability the hustler will *not* correctly guess 1 of the states you have written down is 0.322614+. If we subtract this result from 1 (*certainty*) we will obtain 0.677385+, which is the probability that the hustler *will* guess 1 of the states you have written.

In percentage terms this equals 67.7385+ percent.

Thus, the odds of winning this bet are more than 2 to 1 in favor of the hustler.

Consequently, in every 100 such bets at $20 a round, the hustler will win about 67 of these wagers and lose about 33. Hence the con artist will win 67 × 20 or $1340.

The mark will win 33 × 20 or $660.

The bet clearly favors the hustler.

22

PROPOSITION THAT 1 OF 6 GUESSES
(IRISH COUNTIES) WILL BE CORRECT

"Part of the ten million dollars I spent on gambling, part on booze and part on women. The rest I spent foolishly."—*George Raft*

PROBLEM

Here is another example of how the previous bet may be offered by a con artist.*

A smooth-talking stranger may approach you one evening in a bar and steer the conversation to how much they liked geography while at school. As the minutes pass they may then raise the subject of gambling and how much they enjoy betting, even when the odds do not favor them. As a way of explaining their passion they will take a card from their wallet. On this card all 32 counties that comprise the island of Ireland will be written. To assist those who may not recall the 32 counties of Ireland, I list them here in alphabetical order: Antrim, Armagh, Carlow, Cavan, Clare, Cork, Derry, Donegal, Down, Dublin, Fermanagh, Galway, Kerry, Kildare, Kilkenny, Laois, Leitrim, Limerick, Longford, Louth, Mayo, Meath, Monaghan, Offaly, Roscommon, Sligo, Tipperary, Tyrone, Waterford, Westmeath, Wexford, and Wicklow.

Having showed you the card listing these 32 counties, the friendly stranger will then spin you some story that they are a very good guesser. To emphasize just how good at guessing they are, they will ask you to study the list of the 32 Irish counties on the card. They will then ask you to secretly write the names of any 6 counties on that list on a separate page and will ask you to not let anyone see what you have written.

Let's assume you do as you are asked.

The stranger (or con artist) will then offer you the following bet of $20 at even money odds: they bet you that if they are allowed to make 6 guesses they will be able to guess the name of at least 1 Irish county that you have secretly written down.

You will probably reason as follows: This person has asked me to secretly write the names of 6 of the 32 counties on the island of Ireland. They tell me that if they are allowed 6 guesses, they will be able to correctly guess the name of at least 1 of the Irish counties that I have secretly written. There are 32 counties in Ireland and this stranger is only allowed to make 6 guesses. This bet must favor me. How can it be any other way? I think I will accept this bet and prove this guy does not know much about the theory of probability. I will also pick up a handy $20 in profit if I win.

Of course, the con artist is hoping that your line of reasoning is along the above lines.

The question is this: Should you accept this bet?

---

*Source: Charlie Rice, *Challenge! Fun with Puzzles, Riddles, Word Games and Problems* (Kansas City, MO: Hallmark Cards, 1968), 32. (A slightly different bet.)

SOLUTION

No, you should not accept this bet. As with the previous wagers, the odds in this bet favor the hustler.

Let us first work out the probability that each of the hustler's guesses will *not* successfully match 1 of the 6 counties you have named. The hustler takes their first guess. They guess the name of any county. The probability that their first guess will *not* match any 1 of the counties you have selected is $\frac{26}{32}$. (Remember, you have already chosen 6 counties so they cannot choose 1 of those if their choice is *not* to match 1 of your selected counties.) On their second guess, the probability that their choice will not match 1 of your selected counties is $\frac{25}{31}$. On their third guess, the probability is $\frac{24}{30}$. On their fourth guess, the probability is $\frac{23}{29}$. On their fifth, guess the probability that their choice will not match 1 of your chosen counties is $\frac{22}{28}$. Finally, on their sixth guess, the probability that their choice will not match 1 of your chosen counties is $\frac{21}{27}$.

To obtain the overall probability that none of their guesses matches any 1 of your chosen counties we must multiply those 6 fractions together:

$$\frac{26}{32} \times \frac{25}{31} \times \frac{24}{30} \times \frac{23}{29} \times \frac{22}{28} \times \frac{21}{27} = \frac{165,765,600}{652,458,240} = 0.254063 +$$

Thus, the probability that their 6 guesses will *not* match any 1 of your chosen counties is 0.254063+.

To find the probability that at least 1 of their guesses will match 1 of your chosen counties, we subtract 0.254063+ from 1 (*certainty*) and obtain 0.745936+.

In percentage terms the probability that the con artist *will* match 1 of your selected counties is 74.5936+ percent. The probability they will lose the bet is 25.4063+ percent.

Saying the same thing another way, the odds that the hustler will win this bet are close to 3 to 1 in their favor.

Thus, the odds favor the con artist. They can expect to win 3 out of every 4 wagers over the long term.

It is an attractive bet, for the hustler!

23

PROPOSITION TO SELECT 5 PAT HANDS FROM 25 CARDS

"Let me state what the official IPCC prediction is: sea levels could go up as much as three-quarters of a meter in this century, but there is a reasonable probability it could be much higher than that."—*Steven Chu*

PROBLEM

One evening a stranger may approach you in a bar and turn the conversation toward the subject of strange and remarkable events. They may lead on from this topic to the game of poker.* They will go on to mention the different hands in poker and casually refer to what are known as *pat hands*. *Pat hands* are hands that are unlikely to be improved by dropping cards and selecting new cards in a game of poker. Here are the pat hands from the straight up to a royal flush:

- A *straight* is 5 cards in sequence but not all of 1 suit.
- A *flush* is 5 cards of the 1 suit but not in sequence.
- A *full house* contains 3 cards of the same rank and 2 cards of a different matching rank.
- *Four of a kind* contains 4 cards of the same rank and 1 indifferent card, often referred to as the "kicker."
- A *straight flush* contains 5 cards of the same suit in numerical order.
- A *royal flush* contains the 10, jack, queen, king, and ace, all the same suit.

The fraudster will probably mention the following *pat hands* and the odds of being dealt any 1 of them in 5-card poker: straight (254 to 1), flush (508 to 1), full house (693 to 1), four of a kind (4,164 to 1), straight flush (72,192 to 1), and royal flush (649,740 to 1).

By quoting these odds the hustler is genuinely informing you and other patrons of the bar just how unlikely these hands will be dealt in any poker game.

However, the con artist has an ulterior motive also. They are grooming you for the slaughter. The con artist is preparing to lure you or others into a classic betting proposition that appears to be vastly unlikely but, in fact, drastically favors the hustler.

The hustler will soon get to the bet. They will ask you or one of the other customers to thoroughly shuffle a 52-card deck and then to deal faceup any 25 cards from it.

---

*Source: *Maverick*, season 1, episode 17, "Rope of Cards," January 19, 1958. (It also appeared on *Run for Your Life*, episode 63, September 27, 1967.)

The con artist will then tell you that in a moment they will take these 25 cards and will attempt to make 5 poker *pat hands* from them. In other words, they will attempt to construct 5 hands of cards, where each hand contains 5 cards and where each hand is a straight or better, from the random 25 card hands that you have dealt.

The hustler reminds you that the probability of being dealt *any* pat hand in a game of poker is close to 0.76 percent, which approximately equals $\frac{1}{132}$. (This is correct.) The fraudster will say, however, that they are in a generous mood and prepared to bet $20 at even money odds that they can make 5 *pat hands* from the 25 random cards that were dealt.

Having heard the previously quoted odds concerning the various pat hands you or one of the other customers in the bar may well be tempted to accept this bet. You will reason that the probability of being dealt any *pat hand* in a poker game is low. (The con artist says it is close to $\frac{1}{132}$.) Surely, you will reason, the odds must therefore be tiny that this stranger can construct not 1 but 5 pat hands from the randomly dealt 25 cards.

The question is this: Should you accept the bet?

SOLUTION

Absolutely not!

This is a classic proposition bet that has won many dollars for con artists over the years. The odds of obtaining 5 pat hands from the random collection of 25 cards are dramatically in favor of the hustler.

Let us look at the bet stage by stage.

First the con artist says that the odds are close to $\frac{1}{132}$ that a pat hand will be dealt in any game of poker. Here is how these odds are calculated:

There are 4 possible royal flushes in a game of poker; there are 36 possible straights; there are 624 four of a kind; there are 3,744 full houses; there are 5,108 flushes and 10,200 straights.

The hustler will know that if we add these figures together, we will obtain the total number of pat hands that are possible from five decks: 4 + 36 + 624 + 3744 + 5108 + 10200 = 19,716.

The swindler will probably have been taught by their mentor some years previously that the number of 5-card hands in a deck of 52 cards is 2,598,960. (This equals $52C5$.) They will know that the probability of being dealt a pat hand from 1 deck is *approximately* $\frac{19,716}{2,598,960}$. This equals 0.007586, or approximately 1 divided by 132. In other words, the probability of being dealt a pat hand is about 1 chance in 132. Thus, the fraudster is probably aware that there are about 132 essentially different pat hands that are possible.

The con artist uses these figures to reinforce the belief in the mind of the sucker that it is quite rare to be dealt a pat hand in a game of poker. Since $\frac{1}{132}$ is approximately the probability of being dealt 1 *pat hand*, the probability of being dealt 5 *pat hands* is approximately $(\frac{1}{132})^5$. That is about $1/(4 \times 10^{10})$.

Of course, these figures are used by the con artist so that the mark is inclined to reason that it must be close to impossible to construct 5 pat hands from 25 randomly selected cards.

Of course, this is precisely the kind of thinking the con artist wants the sucker to adopt. It's the first step in luring the mark into accepting the bet.

All the above will probably be pointed out to the sucker by the con artist in order to make the proposition bet attractive.

Now let's get to the core of the matter.

When you deal 25 cards from a shuffled deck the number of arrangements that those 25 cards can take is enormous. The number of ways of choosing 25 cards from 52 is expressed mathematically as $52C25$. To be more precise, there are exactly 477,551,179,875,952 different ways of selecting 25 cards from 52. That is more than 477 *trillion* different selections!

The 25 cards thus selected can be arranged in 25! different ways. This equals 15,511, 210, 043, 330, 985, 984,000,000 different ways. That is about 15 million, million, million, million different ways!

Most lay people find this fact astonishing.

But it is not these astonishing number of ways of arranging the cards alone that sways the odds in favor of the con artist. What makes the odds in this proposition bet favor the swindler can be illustrated by considering the following scenario:

First, let A, B, C, D, and E represent 5 selections or drawings of 5 cards from the 25 cards you randomly selected. The first drawing means 5 cards must first be selected from 25. This is written mathematically as $25C5$. Then 5 cards are chosen from 20. This equals $20C5$. Then 5 cards are selected from 15. This equals $15C5$. Next 5 cards are selected from 10. This equals $10C5$. Finally, one is left with the last 5 cards.

The whole procedure of making the 5 selections of 5 cards is mathematically expressed as $25!/5!^5$. But since the order of the 5 selections is irrelevant, we divide the above expression by 5! This equals $25!/5!^6$, which equals $5.19+ \times 10^{12}$.

The result of our calculations shows that 25 cards can be arranged into 5 hands—each hand consisting of 5 cards—in about 5.2×10^{12} different ways. Saying the same thing another way, 25 cards can be sorted into 5 hands—each hand containing 5 cards—in more than 5 trillion ways!

Consequently, it is extremely likely that it will be possible to assemble at least 5 of the 132 possible *pat hands* since there are 5 trillion poker hands that can be made from the 25 cards.

In other words, it is highly likely that one can construct 5 of the 132 possible pat hands from the 25 cards randomly selected from the deck of 52 cards.

A computer programmer, Connor Stoyle, programed a computer to simulate a *Monte Carlo* simulation of possible pat hands from 25 cards that were randomly dealt from a pack of 52 cards.

The program ran 100,000 trials and reported that 5 pat hands could be arranged 98.0 percent of the time.

Thus, in the long term, the hustler can expect to win this bet about 98 times out of every 100 bets they offer.

The odds in this bet overwhelmingly favor the hustler. It is a fantastic wager with a very counterintuitive solution.

If you want to hold on to your greenbacks, it is best not to accept this bet!

I will point out in passing that it is relatively easy to form a set of 25 cards in which it is *not* possible to form 5 pat hands. For example, consider the following 25 cards, arranged in ascending order within each of their suits:

HEARTS	CLUBS	DIAMONDS	SPADES
2	A	A	A
5	2	4	3
7	3	5	5
Q	6	6	6
	8	7	7
	9	7	8
		10	9
			10

These 25 cards cannot be formed into 5 pat hands. The key card in obtaining a proof of this is the queen of hearts. It's the only queen among the 25 cards. Because there is no king or jack, the queen cannot be part of a straight. Because there are only 3 other hearts, the queen cannot be part of a flush or a straight flush. There are no "4 of a kind" in the 25 cards, so the queen cannot be the fifth card of a "4 of a kind" hand. Because it is the only queen, it cannot be part of a full house. Consequently, the queen has nowhere to go to complete any pat hand. Thus, 5 pat hands *cannot* be formed with these 25 cards.

24

PROPOSITION THAT EXACTLY 2 OF 3 CARDS WILL BE THE SAME SUIT

"Probability is the very guide of life."—*Marcus Tullius Cicero*

PROBLEM

One evening you may be in a bar when a stranger walks in and engages you in friendly conversation.

During conversation, the stranger pulls a deck of cards from their pocket and asks you to examine the deck and then asks you to shuffle them. As you do so, the stranger bets $10—at even money odds—that if you select 3 random cards from the shuffled deck *exactly* 2 of the cards will be of the same suit.

You will probably reason as follows: there are 4 suits in the deck. I am about to select 3 cards. It is surely unlikely that *exactly* 2 of those three cards will be of a similar suit. It must be more likely that all 3 cards are from a different suit.

Therefore, you may well be tempted to accept the bet and in the process believe you may well be teaching this stranger a few basic rules about probability theory.

Of course, this is exactly the type of reasoning that the hustler wants you to adopt.

Should you accept the bet?

SOLUTION

It is best to decline this bet as the odds favor the con artist.

First, we calculate how many ways 3 cards can be chosen from 52. This can be done in 52C3 or 22,100 ways.

Now we calculate how likely it is to select 3 cards from the deck so that *exactly* 2 cards are from the same suit. Since there are 4 suits in a deck there are 4C1 or 4 ways to select the suit. We must draw exactly 2 cards from that suit. Since there are 13 cards in a suit, we find that this can be done in 13C2 or 78 ways.

We now must choose 1 of the other 3 suits. This can be done in 3C1 or 3 ways. Finally, we must choose 1 card from 1 of those 3 suits. This can be done in 13C1 or 13 ways.

Therefore, the probability of selecting 3 cards from the deck so that *precisely* 2 of those cards are of the same suit is:

$$(4C1 \times 13C2 \times 3C1 \times 13C1) / 52C3$$

which equals

$$(4 \times 78 \times 3 \times 13) / 22,100 = 12,168/22,100$$

which equals 0.5505+. In percentage terms this equals 55.05+ percent.

Thus, the probability that *exactly* two of the three cards selected will be of the same suit is 55.05+ percent. Therefore, the probability that exactly two of the three cards will NOT be of a similar suit is 44.94+ percent.

The bet clearly favors the hustler. For every 100 bets at $10 a round, the fraudster can expect to win about $550 and lose $450. That is a profit of $100 in 100 bets, which equals $1 profit per bet.

25

PROPOSITION INVOLVING 2 RED CARDS AND 2 BLACK CARDS

"The 50-50-90 rule: Any time you have a 50-50 chance of getting something right, there's a 90 percent probability you'll get it wrong."—*Andy Rooney*

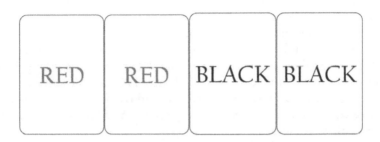

Figure 25.1 2 *red* cards and 2 *black* cards.

PROBLEM

One evening when the sun has gone down you may decide to go to a bar for a few beers. As the evening wears on, a stranger walks up to you and begins a conversation about lotteries. They may even tell you the old joke that lotteries are a tax on people who are bad at mathematics. They may then tell you that they are not good at mathematics, but they do like to gamble now and then, even if the odds do not favor them.

As an example, they take 4 cards—2 cards colored red and 2 cards colored black—from their jeans pocket.* Each card is colored on 1 side only. The other side of each card is blank.

They will then ask you to shuffle the 4 cards and to lay them facedown in a row on a nearby table.

The con artist will emphasize that neither you nor they nor anyone else now knows the positions of the 2 red cards or the 2 black cards on the table.

The hustler then asks you to select any 2 cards and to isolate them from the rest by pushing them—facedown—to the side of the table.

You do so.

Call these 2 cards you pushed to one side card A and card B. They now ask you to turn 1 of those 2 cards faceup.

---

*Source: Martin Gardner, *Wheels, Life and Other Mathematical Amusements* (New York: W. H. Freeman, 1983), chapter 5: Nontransitive Dice and Other Probability Paradoxes.

Let's say you turn up card A and you find it is a card marked red. The hustler now places a $20 bill on the table. They will now bet you $20—at even money odds—that the color of card B is black.

To encourage you to accept the bet, the con artist may add something along the following lines: "We have four cards here. There are 3 possible groups of 2 cards: we can have either 2 red cards, 2 black cards, or 1 red card and 1 black card. In 2 of these 3 groups both cards are similar. So, my friend, the odds favor you. The odds are 2 to 1 in your favor, in fact! But I like to live dangerously. Life is more interesting that way, don't you think! Do you have a similar attitude? You do! Good! Come on pal, what are you waiting for? Put your 20 bucks down and cover my bet. Don't you want the chance of picking up 20 greenbacks?"

The argument put forward by the smooth-talking hustler seems plausible. You may well be tempted to accept the wager. The faceup card on the table is colored red. Your reason tells you that there are 2 chances in 3 that the second card is red also. Therefore, the probability that the second card is black must be 1 chance in 3.

Consequently, you are contemplating placing a $20 bill and betting that the color of the other card (card B) is red.

The question is this: Should you accept the bet?

SOLUTION

This is a classic proposition bet that has emptied the pockets of many marks for many years.

The probability that the second card is marked black is *not* ⅓ but is ⅔. Thus, the odds are 2 to 1 that the second card is black!

The fraudster's reasoning that there are only three possible groups of two cards, Red, Red; Black, Black; Red, Black, is incorrect. Such "reasoning" has tempted many marks to part with their dollars in relation to this bet over the years.

There are 6 ways that 2 of the 4 cards can be selected. We will differentiate between the 2 red cards by designating 1 red card as Red_1 and the second red card as Red_2. Similarly, we designate the 2 black cards as $Black_1$ and $Black_2$. Thus, the 6 ways 2 of the 4 cards can be arranged are as follows:

$$Black_1\ Red_1 \qquad Red_1\ Black_2$$
$$Black_1\ Red_2 \qquad Red_2\ Black_2$$
$$Black_1\ Black_2 \qquad Red_1\ Red_2$$

As you can see, there are 2 cases out of a possible 6 cases in which 2 cards are the same color. Thus, there are 2 chances in 6, which simplify to 1 chance in 3, that the 2 cards you select are the same color. Consequently, there are 2 chances in 3 that the 2 cards you select are each of a different color! In other words, the odds are 2 to 1 in favor that the 2 cards you pick are colored red and black.

Since you initially turned up a red card, the odds are 2 to 1 that the second card is black. The hustler has the edge. They will—over the long term—win this bet 2 out of every 3 times they make the proposition.

Another way of seeing why the odds are 2 to 1 so the second card is of a different color than the first card picked is to consider the following: when you initially select 1 card you know that there are 3 other facedown cards on the table, and that only 1 of them is the same color as the card you have chosen. Therefore, there is only 1 chance in 3 that you will select a card that is the same color as the first card you picked. Consequently, there are 2 chances in 3 that you will select a card that is a different color than the first card you picked. In other words, the odds are 2 to 1 that the second card you select will be a different color from the first card you chose.

It is a fantastic proposition bet.

Don't get caught by it!

26

CLASSIC BETTING PROPOSITION WITH 3 CARDS

"There may be aliens in our Milky Way galaxy, and there are billions of other galaxies. The probability is almost certain that there is life somewhere in space."
—*Buzz Aldrin*

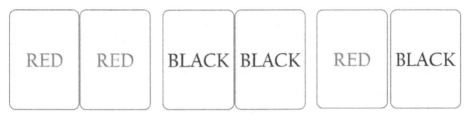

Figure 26.1 Card marked *red* on both sides.

Figure 26.2 Card marked *black* on both sides.

Figure 26.3 Card marked *red* on one side and *black* on the other side.

PROBLEM

You may be at a gathering of friends one evening when someone walks up to you and starts a friendly conversation that eventually leads on to the theme of gambling. As the conversation develops this friendly individual may offer you the following proposition bet: They will show you 3 cards. One card will be marked *red* on both sides. The second card will be marked *black* on both sides. The third card will be marked *red* on one side and *black* on the other side.*

The friendly stranger then gives you the 3 cards to shuffle and asks a third person of your choosing (let's call her Mary) to place the 3 cards in a box.

Both you and the con artist then leave the room. While you are both out of the room Mary is blindfolded by a fourth person and is then asked to take 1 card from the box and place it on the table in full view of all. Mary then takes off her blindfold. She sees that the color of the card on the table is *red*. Mary has no way of knowing what the color is on the other side of the card. In fact, no one knows what the color is on the other side of the card.

You and the hustler then return to the room. You both see that the card on the table is colored *red*.

---

*Source: Warren Weaver, *Lady Luck: The Theory of Probability* (New York: Dover, 1982), chapter 6: Some Problems, 123, 124, 125, 126. A variation of this proposition bet is given in the form of 3 chests.

The con artist turns to you and says, "This is 1 of the 3 cards you shuffled earlier. You can see that the top color of this card is *red*. No one knows what the color of the other side of this card is. However, we do know with certainty that this card is not the card colored *black* on both sides. Therefore, it must be either the card colored *red* on both sides, or it is the card colored *red* on one side and *black* on the other side. It is either one or the other. There is no other possibility. Thus, it is an evens bet. However, I like to instill some excitement into my life. Consequently, I like to gamble every now and then. Now, my friend, I am willing to bet you $20 at 5 to 4 odds in your favor that the other side of that card on the table is marked *red* also."

It seems to you that it is indeed an evens bet. You are contemplating taking the bet, as you believe that you have 1 chance in 2 that you will win. The bet will cost you $20 but if you win, you pick up a handy $25, and get your stake of $20 back.

The question is this: Should you accept the wager?

SOLUTION

Don't accept this classic proposition bet if you want to keep your dollars in your wallet.

The odds are 2 to 1 in favor of the hustler that they will win the bet. To see this, consider the following:

Imagine that the card that is colored *red* on both sides is marked as follows: Red_1 on one side and Red_2 on the second side. The card that is marked *red* on one side and marked *black* on the other side is marked Red_3 and $Black_4$. Then the card that is marked *black* on both sides is marked $Black_5$ on one side and $Black_6$ on the other side. Now the 3 cards are shuffled and placed in a box. One card is then taken from the box and placed on the table by a blindfolded person. We are subsequently informed that the top color of this card is marked *red*.

The hustler then bets you $20 at odds of 5 to 4 in your favor that the color on the other side of the card on the table is *red* also. Is it an even bet, or if not, who does the bet favor?

The bet favors the con artist. As the hustler pointed out, the card on the table cannot be the card that is marked *black* on both sides. It must be 1 of the other 2 cards. But which one is it?

Look at it this way. The card on the table that is marked *red* may be showing the Red_1 side faceup. Or it may be showing the Red_2 side faceup. Or it may be showing the Red_3 side faceup. These are the only 3 possibilities. In 2 of these 3 cases the other side of the card is *red*. In only 1 case of the 3 is the other side of the card colored *black*.

Therefore, the odds favor the hustler when they bet that the other side of the card is *red*. The odds are in fact 2 to 1 in favor of the con artist. In the long term, the swindler can expect to win this bet 2 out of 3 times.

Suppose the swindler—over time—performs 100 of these bets at $20 per bet. They will win about 67 of them, raking in $1,340. For the 33 bets they lose, they pay out 33 times $25, or $825. That still leaves the con artist with a profit of $515. That works out at an *average* profit of $5.15 per bet for the hustler.

It is a very profitable scam for the con artist.

The bet works, of course, in a similar way if the card initially selected from the box is placed so that its top color is *black*.

27

PROBABILITY THAT 2 CARDS OF THE SAME RANK ARE TOGETHER IN A DECK

"Medicine is a science of uncertainty and an art of probability."—*William Osler*

PROBLEM

A con artist may approach you one evening in a crowded room and offer the following proposition bet to you: they will ask you to thoroughly shuffle a 52-card deck, and then bet you $20 at 5-to-1 odds that 2 cards of the same rank are lying next to each other in the deck. (It is important to recognize here that the con artist is not betting that 2 cards of a specific rank are adjacent. They are betting that at least some pair of cards with the same rank will be adjacent.)*

You may be tempted to reason as follows: the number of different permutations of the 52-card deck equals $52 \times 51 \times 50 \times 49 \times \ldots \times 3 \times 2 \times 1$. This number contains 68 digits. Because of the enormous number of permutations, your reason will probably tell you that it must be very unlikely that 2 cards of a similar rank are lying next to each other in a shuffled deck. Therefore, it is very likely that you will consider accepting the bet in the hope of winning a handy $100.

---

*Source: "Combination of Poker Hands": Random Card Shuffling Probabilities, Ask Dr. Math (math forum), http://www.mathforum.org/library/drmath/view.

SOLUTION

You should decline this bet if you want to hold on to your dollars.

Astonishingly the probability that 2 cards of the same rank are lying next to each other in a shuffled deck of 52 cards is more than 95 percent! It is almost certain that the con artist will win this bet.

How is this probability calculated? One way of doing it is to program a computer and use what is known as the *Monte Carlo* method to determine the probability. The *Monte Carlo* method is basically this: a computer is programmed to simulate many shuffles of 52 cards (say 100,000 shuffles). After each simulated shuffle the computer program will examine the resulting mix of the cards within the deck to see if a pair of the same rank is lying beside each other. When such programs are run one finds that at least one pair of the same rank, such as 2 ones, or 2 fives, or 2 kings, are beside each other, about 95.45 percent of the time. (See the online article "The Probability of Cards Meeting after a Shuffle" by Yutaka Nishiyama at https://ijpam.eu/contents/2013-85-5/3/3.pdf.)

This astonishing result can also be obtained *approximately* by mathematical reasoning. For instance, it is easy to see that any card, say the ace of clubs, can only be matched with 3 of its mates. In other words, the ace of clubs cannot be matched with the non-aces in the deck. Saying the same thing another way, the ace of clubs cannot be matched with the other 48 non-ace cards in the deck. Thus, the chance of a match with the ace of clubs is $\frac{3}{51}$. Therefore, the chance that there will *not* be a match is $\frac{48}{51}$.

Similarly, we consider the king of clubs. It cannot be matched with the 48 non-kings in the deck. Or consider the queen of clubs. It cannot be matched with the 48 non-queens in the deck. And so on.

We can go through every card in the deck and *approximately* obtain the same result: the chance of *no* match is $\frac{48}{51}$.

This procedure is not exactly accurate though. Why? Well, for instance, when we consider, say, the 2 of diamonds and say, the 2 of hearts, we find that the 2 of diamonds can be matched with the 2 of hearts. But we then cannot include in our calculations the match of the 2 of hearts with the 2 of diamonds, because we have already counted that match. Thus, there will be discrepancies like these in our calculations. But these discrepancies will be inclined to cancel each other out as we go through every card in the deck.

Thus, we can say that *approximately* the chance of any card *not* matching its mate is $\frac{48}{51}$. Therefore, the approximate chance that all 51 cards will *not* lie beside its mate in the deck is therefore $(\frac{48}{51})^{51}$. This equals 0.0451759+. If we subtract this from 1 (*certainty*) we will obtain the probability that at least 2 cards of the same rank ARE beside each other in the deck.

We find that 1 (*certainty*) − $(\frac{48}{51})^{51}$ equals 0.9545824+. In percentage terms this equals 95.45 percent.

Thus, the *true* probability that 2 cards of a similar rank lie next to each other in a shuffled deck of 52 cards is very close to 95.45+ percent.

Consequently, the hustler is almost certain to win the bet. For every 100 such bets, at $20 per bet, the fraudster will win 20 × 95 or $1900. The 5 bets they lose they pay out 100 × 5 or $500. That still leaves the hustler with a healthy profit of $1400. That is an average profit of $14 per bet for the fraudster.

Sometimes, after performing and winning this bet, the con artist will ask a mark to name the rank of *any* card in the deck. Say the mark names the king. The hustler will then bet, giving odds of 2 to 1 in favor of the mark, that if one shuffles the deck, one will not find a pair of kings adjacent to each other. The probability that a pair of kings (or any specified pair) will *not* be adjacent to each other is approximately 78 percent. Therefore, for every 100 such wagers, the hustler will win this bet 78 times. At $20 per bet, the con artist will rake in $1,560. The 22 times they lose they pay out 40 × 22 or $880. That still leaves the grifter with a profit of $680.

Either of these bets is a handy moneymaker for the con artist.

28

PROPOSITION THAT THERE WILL BE A 10 OR FACE CARD IN 2 CARDS

"The uniformity of the earth's life, more astonishing than its diversity, is account-able by its high probability that we derived originally from some single cell, fertil-ized by a bolt of lightning as the earth cooled."—*Lewis Thomas*

PROBLEM

You may be in a bar one evening enjoying a quiet drink when a friendly stranger approaches you and offers to do a little experiment with a normal deck of cards.*

Having obtained your attention, they will tell you a little story about how the theory of probability has not been verified even by the most respectable mathema-ticians. To support their opinion they will offer the following proposition: they bet you $10 at even money odds that if you pick 2 cards from a thoroughly shuffled 52-card deck, at least 1 of the 2 cards will either be a 10, jack, queen, king, or ace.

Of course, you will recall that there are four 10s, 4 jacks, 4 queens, 4 kings, and 4 aces in a deck. This accounts for 20 cards. You will realize that the bet reduces to this: you are being asked to select 2 facedown cards from a shuffled 52-card deck and the stranger (or con artist) is saying that at least 1 of the cards you will select will be from the group of 20 cards.

You will probably be inclined to reason that the odds are in you favor, because there are 52 cards in a deck, and to win you must avoid a card from a group of 20 cards. Thus, you may well be tempted to accept the bet.

Of course, this is exactly the way the con artist wants you to reason.

Should you agree to the bet?

---

*Source: Oswald Jacoby, *How to Figure the Odds* (Garden City, NY: Doubleday, 1947), chapter 10: Propositions—General.

SOLUTION

No way!

Let the group of 20 cards consisting of four 10s, 4 jacks, 4 queens, 4 kings, and 4 aces be designated Set A.

Let the remaining 32 cards be designated Set B.

The bet is essentially stating that if you select 2 facedown cards from the entire 52-card deck, at least 1 of those 2 cards will be a card that will belong to Set A.

When you pick 2 cards from the 52-card deck the probability that both cards will belong to Set B is:

$$\frac{32}{52} \times \frac{31}{51} = 0.374057+$$

Thus, the probability that at least 1 of the 2 cards will belong to Set A is 1 (*certainty*) − 0.374057+. This equals 0.625943+. In percentage terms this equals 62.5943+.

Therefore, the probability that at least 1 of the 2 cards you choose belongs to Set A is 62.5943+ percent. In other words, there is a 62.5943+ percent probability that 1 of the 2 cards will be either a 10, jack, queen, king, or ace!

Most people are vastly surprised when they first learn of this result.

The odds of winning this bet are almost 2 to 1 in favor of the hustler!

29

PROPOSITION THAT 2 OF 12 PEOPLE
WILL GUESS THE SAME CARD

"In gambling, the many must lose in order that the few may win."
—*George Bernard Shaw*

PROBLEM

You may be at a gathering one evening when a friendly sort of person approaches you and talks about strange events. The conversation will be moving along nicely when the friendly stranger states that they would like to offer a bet to confirm one of their long-held beliefs concerning the frequency of strange and remarkable occurrences.

The stranger puts forward the following simple bet: They ask 12 people at the gathering to secretly write the name of any playing card whatever. The stranger will emphasize that there are 52 different cards in a normal deck. This, they explain, makes it highly unlikely that any 2 of the 12 people will write down the name of the same card.

However the friendly stranger (or con artist) states that they are prepared to bet $20 at even money odds that at least 2 people among the 12 will write the name of the same card.

The other people at the gathering will find this bet intriguing, to say the least. Some of these folk may form the opinion that this stranger has learned their sums all wrong and is making a basic mistake about probability theory. They will reason that only 12 people have been asked to secretly write the name of a playing card. All of these people know that there are 52 different cards in a normal deck. Therefore, it is obvious that the odds must be against the stranger. It is common sense, after all.

Or is it?

The question is this: Should you accept this bet?

SOLUTION

No way!

The odds are in favor of the hustler that they will win this bet.

Let's first calculate the probability that 2 of the 12 people will *not* write down the name of the same playing card.

We see that the first person can choose any card. The probability that they will choose a card is 1 (*certainty*), which we can express as $^{52}/_{52}$. The probability that the second person will *not* match the name of the card of the first person is $^{51}/_{52}$. The probability that the third person will *not* match the name of the cards of the first and second person is $^{50}/_{52}$. The probability that the fourth person will *not* match the name of the cards of the first three persons is $^{49}/_{52}$. And so on. Finally, we reach the twelfth person, where we calculate the probability that the twelfth person will *not* match the name of the cards the previous 11 people chose is $^{41}/_{52}$.

These twelve fractions must be multiplied together to obtain the probability that NONE of the 12 people selected a matching card.

$$\frac{52}{52} \times \frac{51}{52} \times \frac{50}{52} \times \cdots \times \frac{43}{52} \times \frac{42}{52} \times \frac{41}{52} = 0.252908+$$

This product equals 0.252908+.

To find the probability that at least 2 of the 12 persons did choose a matching playing card we subtract 0.252908+ from 1 (*certainty*). This equals 0.747091. In percentage terms this equals 74.7091+ percent.

Thus the probability that at least 2 of the 12 people chose a matching card is 74.7092+ percent. This means that in the long term the odds that at least 2 of 12 people will choose a matching playing card are about 75 to 25, or about 3 to 1 in favor of it happening.

Saying the same thing another way, the odds are—in the long term—that the con artist will win this wager 3 times out of every 4 times they make the bet.

It is a nice little earner for the con artist!

Don't add to their profits by accepting this bet!

30

MATCHING BIRTHDAY PROPOSITION BET

"We didn't push the Russians to intervene, but we knowingly increased the probability that they would."—*Zbigniew Brzezinski*

PROBLEM

One evening a friendly stranger may approach you at a social gathering and engage in conversation with you and a bunch of your pals. They may eventually steer the conversation toward birthdays of famous people.

As they do so, they may make the point that they love to gamble and want to try out an experiment. They count the number of people at the gathering and find that there are, let's say, 30 people.

The stranger then says they are prepared to make the following wager.

They ask all 30 people to secretly write down on a slip of paper their dates of birth (month and day only), then fold each slip and place it in a box. They then bet $50 at even money odds that 2 of those birthdays will match.

On hearing this bet you and the others may be inclined to reason as follows: there are 30 of us here at this gathering. We have all written down our birthdays (day and month only) and this dude bets that 2 of these birthdays will match. But there are 365 different dates in a normal year. The odds must be low that 2 birthdays match. Therefore, the bet favors me and I will accept the wager. It will be a handy way of picking up 50 greenbacks.

Of course, this is the way the hustler wants you to reason, so that you end up accepting the wager.

The question is this: Should you accept the bet?

SOLUTION

No, do not accept this wager!

This is one of the classic proposition bets. With 30 randomly selected people in the room the probability is more than 70 percent that at least 2 birthdays will match! Most people find this result so surprising that they refuse to believe it. (We assume here that the randomly selected people do not include sets of twins or triplets, etc.)

Here's how these probabilities are calculated: The probability a second person does *not* have the same birthday as the first is $364/365$. The probability a third person does *not* have the same birthday as the other 2 persons is $363/365$. The probability a fourth person does *not* have the same birthday as any of the previous 3 people is $362/365$. And so on.

Thus, the probability that any of the 30 people at the gathering does *not* have a matching birthday is:

$$\frac{364}{365} \times \frac{363}{365} \times \frac{362}{365} \cdots \times \frac{337}{365} \times \frac{336}{365} = 0.2936+$$

In percentage terms this equals 29.36+ percent. Thus, there is a 29.36+ percent probability that there will be *no* birthday match among 30 randomly selected persons.

Of course, this means the probability that there *will* be 2 matching birthdays among 30 randomly chosen persons is 1 (*certainty*) – 29.36+ or 70.63+ percent.

If the hustler is doing this bet constantly for groups of 30 randomly chosen people at a time, they will win the bet, in the long term, 7 times out of 10. It is a nice little earner for the con artist.

The truly surprising thing about this bet is that with as few as 23 people chosen at random the odds are about *even* that the birthdays of 2 people among the 23 individuals will match. With 23 people the probability is 0.4927+ that there will *not* be a matching birthday. Therefore, the probability that there will be a matching birthday with 23 people is 0.5072+. This probability is more than evens. With 22 people the probability is 0.5243+ that 2 birthdays will *not* match. Therefore, the probability that 2 birthdays will match in a random group of 22 people is 0.4756+. This probability is less than evens. Thus, 23 people is the critical point for obtaining an evens chance that at least 2 birthdays will match.

This is a very counterintuitive result, and many people find it difficult to believe even when the mathematics involved in the solution is explained to them.

Of course, it would be unwise for a fraudster to propose this sucker bet when there are only 23 people present. If they do, they can expect to win the bet once in every 2 bets over the long term. That is not attractive to the con artist. No hustler in their right mind will operate in this manner! It is much more lucrative to propose this bet when there are substantially more than 23 people present.

With 35 people present the probability of matching birthdays is 81.43+ percent; with 40 people it is 89.12 percent. With 45 people the probability is 94.1 percent and with 50 people it is more than 97 percent probable that there will be a match. With 60 people the probability that 2 birthdays will match is 99.41 percent. With 70 people that probability is 99.92 percent.

With 100 randomly selected people present at a gathering there is only 1 chance in 3,254,690 chances that there will be NO match. Thus, the odds are 3,254,689 chances to 1 that there *will* be a match. Saying the same thing another way, the probability that at least 2 birthdays will match in a random group of 100 people is 0.9999996+. Thus, there is only 1 chance in more than 3 million that there will *not* be a matching birthday with 100 people picked at random! If you propose the bet with 100 randomly selected people, it is almost certain that there will be a matching birthday.

The birthday bet is a true classic among proposition bets.

Don't get taken by it!

31

MATCHING BIRTHDAY PROPOSITION BET WITH A SUBTLE DIFFERENCE

"It is morally wrong to allow suckers to keep their money."—*Canada Bill Jones*

PROBLEM

Having performed the birthday proposition bet, the hustler might now try to make a few more dollars with the following similar bet, which contains a subtle difference, however.

Because the con artist has just picked up a handy $50 on the birthday proposition bet, all 30 people in the group are astonished at the fact that the hustler won the wager. Some of the sharper people in the group will quickly realize that somehow or other the odds favored the con artist.

Of course, the hustler is probably just as sharp as the sharpest in the group and will instinctively feel what the thinking in the group is.

That is why they will spring the following proposition bet on the party of 30 people (which includes you): they will ask you for your specific birthday. They will then wager $100 at even money odds that *none* of the 30 people in the group will share your birthday.

Because the hustler has just won the previous bet with matching birthdays, some in the group, including you, may be inclined to believe that this bet is the same bet as the previous one. Given what has just happened you will probably reason that it is most likely at least 1 other individual in the group of 30 randomly selected people will share your birthday.

The question is this: Should you accept this proposition bet?

SOLUTION

No way!

The previous bet amounted to finding any matching birthday among *any* 2 of 30 randomly selected people. For example, it may have transpired that the fifth and twenty-first people in the group had a matching birthday. Or it may have transpired that the seventh and tenth people in the group had a matching birthday.

This bet however is asking for a match between any of the randomly selected 30 people with *your* birthday. It is a *different* proposition bet from the previous one.

The probability that any 1 of the 30 people's birthday will *not* match your birthday is $(^{364}/_{365})^{30}$. This equals 0.9209+. In percentage terms this equals 92.09 percent.

In other words, the probability is 92.09 percent that there will be *no* match with your specific birthday.

Thus, the probability is 7.9 percent that there *will* be a match.

Thus, it is more than 11 times more likely that there will *not* be a match with *your* birthday and any 1 of the 30 randomly selected people.

The odds overwhelmingly favor the hustler.

Don't accept this bet!

If you do you will almost certainly lose money.

32

COVER THE SPOT

"Is it 10 years, 20, 50, before we reach that tipping point where climate change becomes irreversible? Nobody can know. There's clearly a probability distribution. We need to ensure this planet, and we need to do it quickly."—*Vinod Khosla*

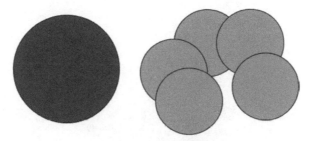

Figure 32.1 Method to successfully cover the entire large red disk.

PROBLEM

A classic proposition bet is that known as *Cover the Spot.** This proposition is usually seen at carnivals in the United States, but now and again a con artist may try the game in a tavern hoping to make a few easy bucks.

The game is easy to play. All one need do is to drop 5 smaller discs from a height of a few inches onto a larger red disc, to completely cover the red spot. The essential point here is that the red disk must be completely covered, so that no tiny part of the red disk can be visible.

The radius of the large disk and the 5 smaller ones varies from operator to operator. Quite often the following dimensions are used. Let the radius of the large disk equal 1. The radius of each of the smaller disks is 0.6180339+.

In such a situation there is just 1 way of arranging the 5 disks to completely cover the large red disk.

The con artist begins by giving several demonstrations. They make it look so easy.

They will then offer you 3 rounds (5 throws per round) for $1. If you succeed in getting the entire red disk covered by the 5 smaller disks you win $1, and you get your stake of $1 back. Usually the prize for winning all 3 rounds is $10 (plus your $1).

Thus, it appears that the odds are in your favor.

The question is this: If this game is proposed to you in a bar or tavern or carnival, should you play?

---

*Source: John Scarne, *Scarne's New Complete Guide to Gambling* (New York: Simon & Schuster, 1974), chapter 25: Cover the Red Spot, 604–607.

SOLUTION

It is best not to play this game if it is offered to you. The operator of Cover the Spot is usually very skilled at the game and makes it appear very easy to cover the large disk by dropping each of the 5 smaller disks from a height of a few inches. However, the mark will find it extremely difficult to replicate the operator's success. If the radius of the large disk is 1 and the radius of each of the 5 smaller disks is 0.6180339, the only way to successfully cover the entire large red disk is shown in figure 32.2.

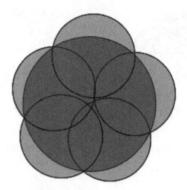

Figure 32.2 Solution to Cover the Spot.

The renowned American gambling expert John Scarne (1903–1985) estimated that, on average, the probability that a mark (that's you!) can drop 1 of the 5 disks from a height of, say, 6 inches onto the correct position it needs to be, so that all 5 disks will completely cover the large disk, is about 1 chance in 3. Therefore, with 5 disks the probability is 1 chance in 3^5, which equals 1 chance in 243.

According to this calculation the odds against the mark successfully covering the entire red disk are 242 to 1! These odds are dreadful from the mark's point of view.

The best bet is not to gamble with this game.

33

PROPOSITION TO MATCH EXACTLY 3 OF 4 CARDS TO CORRESPONDING ENVELOPES

"When you have eliminated the impossible, whatever remains, however improbable, must be the truth."—*Sir Arthur Conan Doyle*

PROBLEM

You may find yourself in a tavern one evening when you see a stranger at the bar having a drink and then taking 4 small cards and 4 small envelopes from their wallet. The stranger places the cards and envelopes in front of them on the bar counter.

Each of the 4 small cards will probably be marked as follows: *a*, *b*, *c*, *d*. Each of the corresponding envelopes is marked A, B, C, D.

You may be intrigued at the thought of what game the stranger is playing and, in a spirit of friendship, approach the apparently likeable person.

In a very short time, the stranger will tell you and the other patrons that they like to gamble, and that they often take a chance when the odds are apparently against them. As an example of the kind of wager they like to make now and then, the smooth-talking stranger will offer the following bet to you.

They ask you to shuffle the four cards marked *a*, *b*, *c*, and *d*, and to place them in a horizontal row, letter-side down, on the bar counter. They now bet you $5 at 10 to 1 odds in your favor that you cannot match up *exactly* 3 of the 4 letter-side-down cards with their corresponding envelopes.

You contemplate the bet for a few moments and calculate that since you need to place 3 cards in 3 correct envelopes there is 1 chance in 3 of you achieving the desired result. Thus, the odds the stranger is offering appear to be generous. You reason that although there are 2 chances in 3 that you will lose the bet, if you win you will be paid odds of 10 to 1. You begin to conclude that maybe this fellow doesn't know as much about probability theory as they would like to believe. Therefore, you are tempted to accept this bet.

The question is this: Should you accept this wager.

SOLUTION

No, do not accept the bet, no matter what odds are offered to you.

The hustler cannot lose in this bet!

You are asked to match up *exactly* 3 of the cards marked *a, b, c, d*, with 3 of the envelopes marked A, B, C, D.

What you are being asked to do is impossible! You cannot *exactly* match 3 of the cards with 3 of the envelopes, because if you do, the 4th card will *also* match its corresponding envelope. In that situation there are 4 cards matching their corresponding envelopes. But the bet stipulated that you match *exactly* 3 of the cards with their corresponding envelopes. Thus, it is impossible to match *exactly* 3 of the cards with their corresponding envelopes.

The hustler is a guaranteed winner with this bet.

If you accept the wager, they will take your money and laugh all the way to the bank!

34

PROPOSITION THAT THERE WILL BE EXACTLY
1 PAIR OUT OF 5 DICE

"It is a truth very certain that when it is not in our power to determine what is true, we ought to follow what is most probable."—*Descartes*

PROBLEM

A stranger may approach you at a social gathering and offer the following bet to you: they bet you $5 at odds of even money that you cannot roll 5 dice with 6 sides, so that the top faces of *exactly* 2 dice will show the same value.*

The con artist will probably point out to you that mathematicians have worked out that the probability of getting at least 1 matching pair is extremely large when rolling 5 dice. (They are correct in this.) The hustler may even mention that the probability that 2 of the 5 dice will *not* have matching faces is $6!/6^5$. This equals 720 divided by 7776, which in turn equals 0.0925925+. Therefore, the probability that at least 2 of the 5 dice *will* match is 1 (*certainty*) − 0.0925925+, which equals 0.9074+. Thus, the probability that at least 2 dice will match is greater than 90 percent. (These figures are correct.)

Then the swindle comes.

The hustler will tell you that if the probability is that high to get *at least* one pair when rolling 5 dice it should be ridiculously easy to get *exactly* one pair.

Therefore, they will say, the odds favor you.

The question is this: Do you accept the bet?

---

*Source: John Scarne, *Scarne's New Complete Guide to Gambling* (New York: Simon & Schuster, 1974), chapter 11: Correct Odds in Dice Games, Using Two, Three, Four or Five Dice, 339.

SOLUTION

The odds in this bet favor the con artist. You may recall that the hustler bet you that you cannot roll *exactly* one pair among the 5 dice. The key word here is the word "exactly."

The probability that you *will* roll *exactly* 1 pair is 25 chances out of 54, which equals 0.46296+. Therefore, the probability you will *not* roll *exactly* one pair is $^{29}/_{54}$, which equals 0.53703+. In other words, in the long run, you will lose this bet 29 times out of every 54 times you accept the bet. This means you will lose the bet 53.703 percent of the time!

How are these probabilities calculated?

Consider the 5 dice. There are a possible $5C2$ or 10 ways of choosing the 2 dice that make up the pair. The values on the faces of the 2 dice can be any 1 of 6 numbers. The other 3 dice can fall in $5 \times 4 \times 3$ or 60 ways. Thus, there are $10 \times 6 \times 5 \times 4 \times 3$ or 3,600 different ways in which *exactly* one pair is rolled among the 5 dice.

There are 6^5 or 7,776 different ways 5 dice can be rolled. Therefore, the probability that you will roll *exactly* one pair is $^{3,600}/_{7,776}$. This simplifies to $^{25}/_{54}$, which equals 0.46296+.

In other words, in the long run, you will roll *exactly* 1 pair just 25 times out of every 54 bets. Thus, you can expect to win the wager 25 times in every 54 bets.

This means, of course, that in the long run the con artist will win the wager 29 times in 54 bets.

The odds are 29 to 25 in favor of the hustler winning the wager.

It is therefore best not to accept this bet.

35

A FAMOUS OLD DICE HUSTLE

"Chance, too, which seems to rush along with slack reins, is bridled and governed by law."—*Boethius*

PROBLEM

The hustler bets at even money odds that the mark (that's you!) will throw both a 6 and 8 with 2 dice *before* they throw two 7s.*

Here is an old dice hustle that is still used today by con artists at street craps games. But the hustle does not just appear on the streets. Occasionally a con artist will pose the bet to an unsuspecting customer in a bar.

Here is how the swindle is usually perpetrated.

The hustler first bets at even money odds that with 2 dice the mark will throw an 8 before they throw a 7. Now the mark, knowing that there are 5 ways to throw 8 and 6 ways to throw 7, is inclined to accept the bet, knowing that they are more likely to throw a 7 rather than an 8, and thereby win the bet. In other words, in this bet, the odds favor the mark.

The mark is feeling good. After a few bets at reasonably low stakes, which the mark tends to win, the hustler switches from betting on 8 to betting on 6. The swindler now bets the mark at even money odds that the mark will throw a 6 before they throw a 7. The mark knows that there are only 5 ways a 6 can be made, compared to the 6 ways a 7 can be made, so once again, the mark knows that the odds favor them (the mark!), and therefore they accept the bet.

After a few small-stake bets, which the mark is inclined to win, the dice hustler now dramatically raises the stakes and performs a subtle swindle. The hustler now bets at *even money* that the mark will throw both a 6 *and* an 8 before they (the mark) throw two 7s. The wager appears to be like the previous 2 bets, so the mark is inclined to accept the bet.

The question is this: If you were offered this wager, should you accept the bet?

---

*Source: Martin Gardner, *Mathematical Magic Show* (Middlesex, UK: Penguin, 1985), chapter 18, Dice (251–262).

SOLUTION

If the mark accepts this bet at odds of even money they have been swindled. Why?

If the hustler had specified which order of the 2 numbers, first a 6 or first an 8, would appear before the mark throws two 7s, the odds would be equivalent to the previous 2 bets offered by the con artist. But the order is not specified by the hustler. Because the order is not specified, the odds have now surprisingly swung in favor of the hustler.

The exact probability in this classic dice hustler's bet is $\frac{4225}{7744}$ or 54.558 percent in favor of the hustler.

Let's take a closer look at the odds involving 2 dice. When 2 dice are rolled there are 36 different possible outcomes. With a throw of 2 dice there are 5 ways to throw an 8, 5 ways to throw a 6, and 6 ways to throw a 7. We need only consider these 16 possibilities. All other outcomes—in relation to this swindle—are irrelevant.

There are 2 cases to consider. The first case is where one rolls *either* a 6 *or* an 8 and then the *other* number before rolling a 7.

The second case is where a 7 is rolled first and then one tries to roll either a 6 and an 8 *or* an 8 and a 6 before rolling a second 7.

Consider the first case. Overall, we are considering 16 cases. On your first roll there is $\frac{10}{16}$ chances you will obtain the number you need, and $\frac{6}{16}$ chances you will obtain a 7. Assume for the moment you have obtained the number you need—say a 6.

When you now roll the 2 dice, you must avoid getting two 7s. The chances of getting two 7s are $\frac{6}{11} \times \frac{6}{11} = \frac{36}{121}$. (The 11 in the denominator is there because we discard the 5 chances in 16 probability of getting an 8.) The chances of NOT getting two 7s are therefore $\frac{85}{121}$. To get to this point we had to overcome odds of $\frac{10}{16}$ (recall that to get to this point we had to first roll either a 6 or an 8). Therefore, to win in this first case the odds are $(\frac{85}{121}) \times (\frac{10}{16}) = \frac{425}{968}$.

Consider now the second case. This is the situation in which an initial 7 has been rolled. One now needs to roll either a 6 *or* an 8 and then roll the *other* specific number before rolling a second 7. There are $\frac{10}{16}$ chances of rolling either a 6 or an 8. As soon as one of these numbers is rolled, the chances of getting the other number and winning are $\frac{5}{11}$. Thus, the chances of getting this far in this second case are $(\frac{10}{16}) \times (\frac{5}{11}) = \frac{25}{88}$. But to get to this point one had to initially throw a 7. The odds of getting this are $\frac{6}{16}$. Thus, the chances of initially throwing a 7 followed by either a 6 or an 8 or an 8 and a 6 are $(\frac{25}{88}) \times (\frac{6}{16}) = \frac{75}{704}$.

We now add together the probabilities of the 2 cases: $\frac{425}{968} + \frac{75}{704}$. The result is the probability that the mark will throw both a 6 *and* an 8 before they throw two 7s.

We find that $\frac{425}{968} + \frac{75}{704} = \frac{4225}{7744}$. This fraction equals 54.558+ percent. This is the probability that the mark will throw both a 6 and an 8 *before* they throw two 7s. This is what the hustler is betting that you will do. In other words, the odds in this bet favor the con artist.

This means the dice hustler will win the bet about 54.5583+ percent of the time.

Summing up, we find that 54.5583+ percent of the rolls of 2 dice will result in a sum of either a 6 or an 8 or an 8 and a 6. Just 45.4417+ percent of the rolls will result in two 7s.

The odds favor the hustler by about 55 to 45, which equals 11 to 9! Consequently, over the long term, the con artist can expect to win 11 of every 20 bets they make.

This a famous dice swindle that is still perpetrated to this day.

If you are offered this bet, look the other way!

36

PROPOSITION BET INVOLVING GUESSING
THE NUMBERS 0 TO 99

"Fate laughs at probabilities."—*Edward Bulwer-Lytton*

PROBLEM

You may be in a bar one evening when a stranger brings up the subject of paranormal activity and may claim that some people have extraordinary "powers" at guessing numbers. As an example of this phenomenon they will offer to demonstrate their own "powers" if someone is prepared to take a little bet from them.*

Let us assume you take them up on their offer.

The stranger will ask the bartender for a telephone directory. They will then ask you to open the directory to any page and to draw a circle around any 20 consecutive numbers.

Because you are a nice, friendly person, you do as you are asked.

The stranger then proposes the following wager: they bet $20 at even money odds that at least 2 of those 20 consecutive telephone numbers end with the same two digits.

The stranger lays a $20 note on a table and asks you to do the same.

On hearing this proposition, you will probably reason as follows: this stranger has asked me to open this telephone directory to a random page and to encircle any 20 consecutive telephone numbers on that page. They then tell me that at least 2 of those numbers will end in the same two digits! Not only that, but they are betting me $20 at even money odds that this will be the case. I believe this person must have this wrong, because the odds must be against them with this wager. Everyone knows that there are 100 two-digit numbers from 00 to 99. Yet we are only checking the end two digits of 20 consecutive numbers. It's logical that 2 of the 20 numbers will not end with the same 2 digits. It's common sense that this stranger has got the odds wrong on this! I think I will accept their bet and pick up a handy $20.

You then take a $20 bill from your wallet and place it on the table.

The question is this: Should you accept this bet?

---

*Source: Oswald Jacoby, *How to Figure the Odds* (Garden City, NY: Doubleday, 1947), chapter 10: Propositions—General.

SOLUTION

This type of bet is similar to the matching birthday problem and is solved in a similar way.

There are 20 consecutive telephone numbers in the directory. The last 2 digits of each telephone number can be any one of the 2-digit numbers from 00 to 99. There are 100 such numbers.

The probability that the second number does NOT match the first number on the list is $^{99}/_{100}$. The probability that the third number does not match the first or second number is $^{98}/_{100}$. The probability that the fourth number does not match the first 3 numbers on the list is $^{97}/_{100}$. And so, the calculations proceed, each time calculating the probability that each successive number does not match any of the preceding numbers. The calculations look like this:

$$\frac{99}{100} \times \frac{98}{100} \times \frac{97}{100} \cdots \times \frac{82}{100} \times \frac{81}{100} = 0.13039+$$

By the time the twentieth number is reached the probability that all 20 numbers end with two *different* digits equals 0.1303+.

Therefore, the probability that at least 2 numbers end with two *similar* digits is 0.1303+ subtracted from 1 (*certainty*). This equals 0.8696+. In percentage terms this equals 86.96+ percent.

Therefore, the probability that at least 2 of the 20 numbers *will* end with 2 *similar* digits is nearly 7 times more likely than any of the 2 numbers *not* ending with *similar* digits.

Consequently, the odds are nearly 7 to 1 that the con artist will win this wager.

It is a fantastic proposition bet.

If you agree to this wager, you are almost certainly going to lose!

37

PROPOSITION OF GUESSING A SPECIFIC 2-DIGIT NUMBER

"If you're playing a poker game and you look around the table and can't tell who the sucker is, it's you."—*Paul Newman*

PROBLEM

Having just taken 20 greenbacks off you with the previous bet the hustler may well spring the following wager on you if you happen to be walking with them down the sidewalk of a busy street.* They will say something along these lines: "Name any 2 digits, my friend. What's that? Oh, I see. You named 5 and 7. Well, I'll tell you something, my friend. I bet you $20 that if we check the tag number of the first 50 cars we both see, we will *not* see a license plate ending in the number 57."

Having been stung for $20 on the previous wager you will be inclined to think that this is a similar bet and that consequently the odds favor you on this occasion.

Of course, this is exactly what the swindler wants you to think.

The question is this: Should you accept this bet?

---

*Source: Oswald Jacoby, *How to Figure the Odds* (Garden City, NY: Doubleday, 1947), chapter 10: Propositions—General.

SOLUTION

No way!

Although this wager appears to be like the previous bet, it has one subtle but essential difference. In this bet you have chosen a 2-digit number, in this case 57. The con artist is now betting you that of the first 50 cars you both see not one of them will have a license plate that ends with 57.

Let's work out the probabilities of this classic proposition bet.

The probability that the first car you both see does *not* end with 57 is $^{99}\!/_{100}$. The *same* probability applies to the second car you meet, and to the third car, and to the fourth car. And so on.

In checking the first 20 cars that you meet the probability that a car's license plate will *not* end in 57 is $(^{99}\!/_{100})^{20}$, which equals 81.79 percent.

In checking the first 40 cars you will meet the probability that a car's license plate will *not* end with 57 is $(^{99}\!/_{100})^{40}$, which equals 66.89 percent.

In checking the first 50 cars you will meet, the probability that a car's license plate will *not* end with 57 is $(^{99}\!/_{100})^{50}$, which equals 60.50 percent.

Consequently, this means that in checking the first 50 cars that you both meet, the probability that the 2 end digits of the license plate will *not* match your specific 2-digit number is 60.50 percent. Therefore, the probability that the 2 end digits of the license plate *will* match your two-digit number is 39.50 percent.

The odds are therefore very close to 60 to 40, or 3 to 2 in favor of the con artist's assertion that there will *not* be a match.

Consequently, in the long term the con artist will win 3 out of every 5 bets made. That means you will lose 2 out of every 5 such bets.

To get the odds down to *even* you would have to check the first 69 cars you see. In other words, $(^{99}\!/_{100})^{69}$ equals 0.4998+.

Thus, if the con artist had offered the bet that you both check the end 2 digits on the license plate of the first 69 cars you would meet, it would be an even-money bet.

But, of course, hustlers do not operate in such a manner. (They would go out of business quickly if they did!) The con artist stipulated that you both check the first 50 cars that you both meet. As we have seen the odds in that case are 3 to 2 in favor of the hustler.

That gives the con artist a very profitable edge. Do not accept this bet, because the odds of your winning are against you.

38

A SURPRISING HUSTLE WITH PLAYING CARDS

"Only those who risk going too far can possibly find out how far one can go."
—*T. S. Eliot*

PROBLEM

Here is a surprising hustle with playing cards. You may be in a bar one evening when a stranger offers the following proposition: the hustler hands a deck of 52 cards to you and asks you to check that the deck is a normal deck of cards. When you have finished examining the deck and you're satisfied that the deck is a normal one, the hustler asks you to blindfold them.

When this is done the hustler asks you to reverse as many cards in the deck as you wish, and to inform them how many cards you have reversed. Let's assume you reverse 15 of the cards in the 52-card deck and then inform the hustler of this. In other words, the 15 cards that are reversed are now facing up in the deck while the other 37 cards are facing down.

The hustler now asks you to shuffle the deck of cards as often you wish. The reversed cards are now thoroughly mixed in the deck.

The shuffled deck is now placed down on the table in front of the blindfolded hustler. The con artist now bets $50 at odds of 10 to 1 in your favor that they will—with the blindfold still on—divide the deck of 52 cards into two piles, so that the number of faceup cards in both piles will be equal.

You will probably be astonished that the con artist is offering such a bet. There appears to be no way that the hustler can know where the 15 faceup cards are in the deck. No one else knows—including you—the positions of the reversed cards either! Thus, it appears impossible for the con artist to divide the deck into two piles so that there will be an equal number of faceup cards in each pile.

With their blindfold still on the con artist reaches into one of their pockets, takes out a $50 bill, and places it on a nearby table. They ask one of the onlookers in the bar to place their 50 greenbacks on top of the $50 bill. The hustler repeats their bet, saying, "There is my 50 bucks. I say that I can divide the deck into 2 piles so that the number of faceup cards in both piles is equal. I am offering odds of 10 to 1 in favor of the challenger that I can successfully do this."

You instinctively feel they have got the odds on this bet entirely wrong. Because the bet seems very favorable to you, you are contemplating accepting the wager in the hope of perhaps collecting $500, plus your stake of $50 back.

You take a $50 bill from your wallet and are wondering if you should place it on top of the hustler's bill on the table.

Should you accept the bet?

SOLUTION

Absolutely not!

The hustler does not need to know where the reversed cards in the deck are to win the wager.

To win the bet the con artist simply divides the 52 cards into 2 piles (call them pile A and pile B) so that there are 15 cards in pile A and 37 cards in pile B.

The con artist now simply flips pile A over. Surprisingly the number of faceup cards in each pile will now be equal. To your astonishment, the hustler has now won the bet.

Why does this procedure work?

When the con artist divides the deck into two piles of 15 and 37 cards the hustler obviously does not know how many faceup cards are in either pile.

However, they do know (because you told them) that 15 cards in the deck were reversed by you earlier. Consequently, when the con artist divides the deck into 2 piles, the hustler can reason as follows: let the number of faceup cards in pile A equal n, where n is less than or equal to 15. The number of faceup cards in pile B now must be equal to $15 - n$.

At this point the hustler can easily reason that the number of *facedown* cards in pile A also has to be $15 - n$. Thus, by simply flipping pile A the number of facedown cards in that pile ($15 - n$) become faceup. Consequently, the number of faceup cards in both piles is now equal.

The con artist wins the bet!

An example will make the procedure clear.

You have informed the con artist that 15 cards in the deck of 52 were reversed by you. In other words, you have reversed 15 cards so that they are now facing up in the deck while the other 37 cards are facing down.

The hustler now asks you to shuffle the deck of cards as often as you wish. The reversed cards are now thoroughly mixed in the deck.

The con artist now divides the deck into 2 piles, by placing 15 cards into one pile (call that pile A), and the remaining 37 cards into the second pile (call that pile B).

Suppose—for the sake of argument—that of the 15 cards in pile A there are 6 cards that are faceup. This means that the number of cards that are faceup in pile B must be 9, since the total number of cards that are faceup in the entire deck is 15. It also means that the number of cards that are *facedown* in pile A is 9. Thus, by simply flipping pile A the number of cards that are *facedown* (9) become faceup. Consequently, both piles contain the same number of faceup cards. Thus, the hustler wins the bet.

Provided that the deck is divided so that the number of cards in 1 of the 2 piles equals the total number of cards that were reversed, the flipping procedure described here guarantees that the number of cards that are faceup in both piles is equal.

Consequently, the hustler can offer any odds they wish to the mark to attract their interest in accepting the bet. Of course, the con artist offers odds of 10 to 1 or somewhere in that range (knowing that their stake is safe) in order to lure the mark into accepting the bet. At the end of the day it does not matter what odds are offered by the con artist to the mark, because the hustler is *guaranteed* to win the bet.

The hustler is the one who will walk away with the winnings—which are your dollars—in this bet.

39

BET INVOLVING 4 NONTRANSITIVE DICE

"When your opponent's sitting there holding all the aces, there's only one thing to do: kick over the table."—*Dean Martin*

PROBLEM

You may be socializing with friends one evening when a stranger approaches you. The stranger produces 4 6-sided dice that are colored *blue*, *yellow*, *black*, and *green*. They will point out that these dice are not numbered in the normal way. In fact, they will show you that the 4 dice are numbered as follows:

Blue die	4, 4, 4, 4, 0, 0
Yellow die	3, 3, 3, 3, 3, 3
Black die	6, 6, 2, 2, 2, 2
Green die	5, 5, 5, 1, 1, 1

The stranger tells you they would like to play a little dice game with you. The game is as follows.*

You will first choose any 1 of the 4 dice. You do so. They will then choose 1 die. They emphasize you have the first choice. They will then ask you to roll your selected die. When you have done so they will roll their die. The player who gets the highest number on either of the two dice wins.

They offer to play 21 of these dice games with you. Each time you play you get to choose first which die you want to play with. Only when you have made your choice does the stranger get a chance to choose the die they wish to play with.

The stranger tells you that whoever wins the most of the 21 games wins the bet. They propose to play for $20. Thus, whoever wins the most of 21 games wins $20.

The game seems fair. In fact, it seems you have the advantage in each game, because each time you play you get to choose first which die you want to play with.

The question is: should you choose to play this game?

---

*Source: Martin Gardner, "Mathematical Games: The Paradox of the Nontransitive Dice and the Elusive Principle of Indifference," *Scientific American* 223 (December 1970): 110–14.

SOLUTION

No way!

The bet favors the hustler!

Consider carefully the numbers on the 4 dice (see the previous page).

Each die on this list is beaten by the previous die with a probability of ⅔.

Let's see how this works in practice.

Suppose you pick the *yellow die*. Then the hustler will choose the *blue die*. Because of the way the dice are numbered they will beat you on average in 4 of the 6 possible ways each die can fall.

Suppose you pick the *black die*. Then the hustler will choose the *yellow die*. Once again because of the way the dice are numbered they will beat you on average in 4 of the 6 possible ways each die can fall.

Suppose you choose the *green die*. The hustler will then choose the *black die*. They will again beat you on average in 4 of the 6 ways each die can fall.

Suppose you choose the *blue die*. The hustler will then choose the *green die*. Once again, the hustler will win on average 4 of the 6 possible outcomes in which each die can fall.

Thus, no matter which die you choose the con artist can choose another die that will beat yours on average 4 times in every 6 throws of the dice.

This means that the hustler has on average a probability of 66.6666+ percent of winning the bet. Consequently, if you agree to play 21 games against the con artist, the hustler can expect to win about 14 of the 21 games. Saying the same thing another way, the odds are about 2 to 1 in favor of the hustler winning the bet.

This property of the four dice is known as a nontransitive property. This layout of the four dice was discovered by the United States statistician Bradley Effron (1938–). Consequently, these dice are often called *Effron Dice*.

40

PROPOSITION THAT 2 OF 4 CARDS
ARE FROM THE SAME SUIT

"The roulette table pays nobody, except him that keeps it. Nevertheless a passion
for gaming is common, though a passion for keeping roulette tables is unknown."
—*George Bernard Shaw*

PROBLEM

The hustler asks you to shuffle a normal deck and to then turn up the top 4 cards.
The con artist then bets you $10—giving you odds of 5 to 1 in your favor—that *at
least* 2 of the 4 cards you turn up will be of the same suit.*

One evening at a social gathering a stranger may engage you with small talk for
a few minutes. Very soon, however, the hustler will bring the conversation around
to gambling.

The con artist will ask you to shuffle a normal deck of 52 cards and to then turn
up the top 4 cards. They will then bet you $10—giving the apparently very generous
odds of 5 to 1 in your favor—that at least 2 of the cards you turn up will be of the
same suit.

The hustler will point out that according to the laws of probability it is to be
expected that the 4 cards that are turned up will all be of a different suit. But they
emphasize that they are getting strong vibes that the normal expectation may not
happen on this occasion. Consequently, they are prepared to give you generous odds
on this little experiment.

The hustler's explanation appears plausible to the uninitiated. That leaves you
wondering what to do. If you bet $10 and lose, you are down $10. But you tell
yourself that you can afford to lose your stake of 10 bucks. On the other hand, if
you win, you get 50 greenbacks, plus the $10 you staked!

The bet appears very tempting.

The question is this: Should you accept the wager?

---

*Source: Oswald Jacoby, *How to Figure the Odds* (Garden City, NY: Doubleday, 1947),
chapter 10: Propositions—General.

SOLUTION

No way!

Consider the normal deck of 52 cards. Let us now look at the probability of turning up 4 cards where *none* of the cards are of a similar suit.

The first card turned up can be any 1 of the 52 cards in the deck. It must belong to some suit. Therefore, the probability it belongs to some suit is 1 (*certainty*). We can express that probability as $\frac{52}{52}$.

If the second card is to be of a different suit than the first card turned up, it must be 1 of the other 39 cards. Therefore, the probability of the second card turned up being a card of 1 of the other 3 suits is $\frac{39}{51}$.

Similarly, if the third card is to be of a different suit than the first 2 cards turned up, it must be 1 of the other 26 cards. Therefore, the probability of the third card turned up being a card of the remaining 2 suits is $\frac{26}{50}$.

By similar reasoning we deduce that the fourth card turned up must be 1 of the remaining 13 cards of the suit that has not yet been chosen. Therefore, the probability of the fourth card turned up being a card of the last remaining suit is $\frac{13}{49}$.

We now multiply these 4 fractions together to obtain the probability that the top 4 cards turned up are all from different suits.

Our calculations look like this:

$$\frac{52}{52} \times \frac{39}{51} \times \frac{26}{50} \times \frac{13}{49} = \frac{685,464}{6,497,400} = 0.1054+$$

Therefore, the probability that the top 4 cards are all different suits is 0.1054+.

If we subtract that result from 1 (*certainty*) we will obtain the probability that *at least* 2 of the 4 cards turned up are of a similar suit: 1 − 0.1054+ = 0.8945+.

Therefore, when 4 cards are turned up on top of a shuffled deck the probability that *at least* 2 of them will be of a similar suit is 0.8945+. In percentage terms this equals 89.45+ percent.

Thus, the odds are nearly 9 to 1 in favor of the hustler!

For every 100 bets, the con artist can expect to win about 90 of them. At $10 a go this will yield them $900. The 10 bets they lose—at odds of 5 to 1 in favor of the sucker—will cost them 10 times $50, or $500.

That still leaves the hustler with a healthy profit of $400 for every 100 bets! That works out at an *average* profit of $4 per wager.

Don't accept this proposition bet if you wish to keep your dollars in your wallet!

41

THE DOLLAR BILL BET

"Once is happenstance. Twice is coincidence. Three times is enemy action."
—*James Bond*

PROBLEM

You may be at a gathering one evening when a neatly dressed stranger walks up to you and offers you the following proposition bet: They will ask you to take a dollar bill from your wallet. They will point out that the dollar bill contains 8 digits from a possible 10 digits (0 through to 9).

The stranger will then bet you $10 that you cannot guess 3 of the 8 digits on your dollar bill. (For the purposes of the puzzle we will assume that there are an equal number of dollar bills with all possible 8-digit combinations in circulation.)*

The question is this: Should you accept the bet?

---

*Source: Michael Shackleford, A.S.A. (also known as the *Wizard of Odds*), "Problem 178: Serial Number Problem," mathproblems.info, www.mathproblems.info/working.php.

SOLUTION

Absolutely not! If you do accept this bet you are most likely going to lose your stake.

The *exact* probability that you will be able to correctly guess 3 digits on the dollar bill is 0.15426684. That means the probability you will *not* be able to correctly guess the 3 digits on the dollar bill is 0.84573316. In percentage terms this equals 84.57+ percent.

This means it is more than 5 times more likely that you will *not* correctly guess 3 digits on a dollar bill!

The method of calculation used in obtaining the *exact* probability of the appearance of your 3 chosen digits on a dollar bill requires some careful reasoning.

Let us assume that a stranger offers you this proposition bet. Suppose you name the 3 digits 0, 1, and 2. How does one calculate the probability that 0, 1, and 2 will appear on the dollar bill?

First, we note that the digits on the serial number of a dollar bill can be arranged in 10^8 different ways, because there are 10 possible digits from 0 to 9, and there are 8 different positions they can occupy in the serial number.

Second, we note that the number of ways the 7 digits from 3 to 9 can occur in the serial number is 7^8, because there are 7 digits and 8 places in which they could appear in the serial number. Call the result (7^8) Result A.

The number of ways the serial number can contain any 1 of the 3 selected digits is $8^8 - 7^8$. Therefore, the number of ways the serial number can contain just 1 of the 3 chosen digits is $3 \times (8^8 - 7^8)$. Call this Result B. Therefore Result A + Result B equals the number of serial numbers containing just 1 of your chosen numbers. This equals $7^8 + 3 \times (8^8 - 7^8)$.

We calculate that the number of ways the serial number can omit 0 is 9^8. Some of these 9^8 serial numbers, however, will not contain the digits 1 and 2. The number of serial numbers with 1 but not 2, is $8^8 - 7^8$. Also the number of serial numbers containing a 2 but not 1 is $8^8 - 7^8$. Therefore the number of ways the serial number contains at least one 1 and at least one 2 but not 0 is $9^8 - 2 \times (8^8 - 7^8)$.

We calculated earlier that the number of ways the serial number contains *only* the digits from 3 to 9 is 7^8. Thus 7^8 must be subtracted from $(9^8 - 2 \times (8^8 - 7^8))$ to give $(9^8 - 2 \times (8^8 - 7^8)) - 7^8$. (This is the number of ways the serial number contains at least one 1 and at least one 2, but not 0, and *none* of the serial numbers that contain *only* digits from 3 to 9.)

This result has to be multiplied by 3, to account for the fact that the omitted 0 in the previous calculation can be any 1 of the 3 chosen numbers.

This gives the equation $3 \times (9^8 - 2 \times (8^8 - 7^8)) - 7^8$. Call this Result C.

Therefore the probability that the serial number will *not* contain all 3 selected numbers, 0, 1 or 2, is equal to Result A + Result B + Result C, all divided by 10^8.

This equals $(7^8 + 3 \times (8^8 - 7^8)) + 3 \times (9^8 - 2 \times (8^8 - 7^8)) - 7^8) / 10^8$ and simplifies to $(7^8 + 3 \times 9^8 - 3 \times 8^8) / 10^8$ or 0.84573316.

If we subtract this from 1 (*certainty*) we will obtain the probability that the serial number *will* contain all 3 selected digits. Thus $1 - 0.84573316 = 0.15426684$ and the probability that the serial number will contain the 3 selected digits is 0.15426684.

As you can see the odds in this famous proposition bet strongly favor the hustler. This proposition bet has appeared in the recreational mathematics and magic literature over the years. It is well known to serious gamblers who like to spring a proposition bet on an unsuspecting mark from time to time.

42

PROPOSITION OF CHOOSING THE QUEEN BEFORE A KING OR JACK

"The safest way to double your money is to fold it over once and put it in your pocket."—*Kin Hubbard*

PROBLEM

You may be in a bar one evening when an attractive person walks in, orders a drink, and takes out a deck of cards. The stranger will sit at the bar, shuffling the deck and apparently examining the cards as they slowly drink their beer. It is only a matter of time before another customer will happily approach the attractive stranger.

We will assume for the sake of argument that *you* are the customer who approaches the newcomer.

Eventually the stranger will give you 13 cards of the hearts suit. They will ask you to thoroughly shuffle the 13 cards and to place them facedown on the bar counter. No one now knows how these 13 cards are distributed.

They offer you a little proposition bet. They bet you $10 at even money odds to choose just 1 card at a time from the 13 facedown cards on the counter, and to place your chosen card faceup in a separate pile. If you choose the queen of hearts *before* you choose the king or jack, you win. But if you choose the king or the jack *before* you choose the queen, you lose.

What, they ask, could be simpler?

The stranger will probably point out to you that there are only 2 of the 13 cards that you must avoid choosing: the king and jack. You can choose any other cards. All you need do is choose the queen *before* you choose the king or jack. They will tell you the odds overwhelmingly favor you.

You contemplate the bet. It seems to you that the odds do, in fact, favor you. There are 13 cards on the table and all you need do is avoid choosing 2 cards: the king and jack. If you can avoid choosing those 2 cards you will profit to the tune of $10. Thus, maybe you should accept this bet. It might be a handy way of picking up 10 greenbacks.

The question is this: Should you accept the wager?

SOLUTION

It is best not to accept this bet. The odds favor the con artist!

Consider the entire 13 cards on the bar counter. Of these 13 cards only 3 of them interest you: the queen, king, and jack. All the other cards are red herrings, put there by the con artist to distract you from calculating the true probabilities of this swindle.

You start choosing the cards one at a time. You can pick any 1 of the other 10 cards and you are still in the game. The game will only end when you choose either the king or jack *before* you pick the queen, in which case you lose, or when you pick the queen *before* you choose the king or jack, in which case you win.

Thus, the game reduces to this: is it more probable that you will pick the king or jack *before* you choose the queen?

The 3 cards that interest you are the queen of hearts, king of hearts, and jack of hearts. There are just 6 ways these 3 cards can be chosen.

FIRST CHOICE	SECOND CHOICE	THIRD CHOICE
queen	king	jack
queen	jack	king
king	queen	jack
king	jack	queen
jack	king	queen
jack	queen	king

One finds that there are just 2 ways of choosing the queen *before* choosing the king or jack.

There are 4 ways of picking either the king or the jack *before* one chooses the queen.

Thus, the odds are 4 to 2 against you that you will choose a queen first. These odds simplify to 2 to 1 against you and in favor of the hustler.

In the long term the hustler can expect to win this wager 2 out of every 3 bets.

It is a great bet, for the hustler!

43

THE MAGIC SQUARE PROPOSITION

"The house doesn't beat the player. It just gives him the opportunity to beat himself."—*Nicholas Dandolos*

PROBLEM

You may be with a group of friends one evening when a stranger approaches and asks you and your pals if you have ever seen a magic square. The stranger will explain that a magic square is a square array of numbers arranged in such a way that each horizontal row, each vertical row, and each of the 2 diagonal rows all have equal sums. The smallest magic square contains 3 horizontal and 3 vertical rows.*

If you respond by saying you have never seen a magic square, the stranger will illustrate one on a separate page. If you say you have seen a magic square before, the stranger will still insist on showing you an example of one and perhaps mention that they will show you some arcane property of that magic square.

The hustler will then take 9 cards, from ace through 9, and all from 1 suit, from their wallet, and lay them down as follows:

8	ace	6
3	5	7
4	9	2

You will notice that the digits in each horizontal row, each vertical column, and each of the 2 diagonal rows sum to 15, if we give a value of 1 to the ace.

Reading across the horizontal rows, you will see that the 9 cards are divided into 3 sets, each set containing 3 cards. Call each of these Sets A, B, and C. The con artist will ensure that the 3 sets of cards will all have different colored backs. For example, the 3 cards in Set A will have red backs; the 3 cards in Set B will have blue backs, and the 3 cards in Set C will have gold backs.

Having informed you of one of the unusual properties of the magic square (to make the conversation interesting), the fraudster will ask you to shuffle the 3 cards in Set A and to place each card facedown across the top row.

The hustler asks you to do the same thing with the 3 cards in Set B (the middle row) and with the 3 cards in Set C (the bottom row).

No one now knows how the cards are distributed in each of the 3 horizontal rows.

---

*Source: Martin Gardner, *Time Travel and Other Mathematical Bewilderments* (New York: W. H. Freeman, 1988), chapter 5: Nontransitive Paradoxes.

The con artist now asks you to draw a card from any 1 of the 3 horizontal rows. They will emphasize that you have a free choice. Before you choose your card however the hustler says that they will draw a card—after you have drawn yours—from 1 of the other 2 sets.

Thus, if you draw a card from the top row, the hustler will draw a card from either the middle row or the bottom row. They say they are prepared to bet $10 at even money odds that the card they draw will be of a higher value than yours. (The ace represents 1 in this bet.)

The con artist places a $10 bill on a table and asks you to cover it. In other words, they ask you to place your $10 bill on the table and agree to the bet.

The question is this: Should you accept the wager?

SOLUTION

No, you should not accept the bet.

No matter what row you draw a card from, the hustler can choose another row, which gives them odds of 5 to 4 in their favor that they will choose a card that has a higher value than your card!

Here's the thing. When you have shuffled each of the 3 rows of cards no one knows how the cards are distributed in each row. But the hustler does not need to know this information.

We know that each of the 3 rows of cards contains the following cards:

ace	6	8
3	5	7
2	4	9

Suppose you draw a card from the middle row (Set B). Both you and the hustler know that you must draw either the 3, 5, or 7.

The hustler now draws a card from the top row (Set A). If they draw 8, they have 3 chances in 3 of beating your card. If they draw 6, they have 2 chances in 3 of beating your card. If they draw ace, they have 0 chances in 3 of beating your card. Consequently, they have 5 chances in 9 of beating your card.

Suppose however you had drawn a card from the bottom row (Set C). Both you and the hustler now know that you have either the 2, 4, or 9.

The con artist now draws a card from the middle row (Set B). If the hustler draws a 7, they have 2 chances in 3 of beating your card. If they draw a 5, they have 2 chances in 3 of beating your card. If they draw a 3, they have 1 chance in 3 of beating your card. Consequently, they have 5 chances in 9 of beating your card.

Suppose however you had drawn a card from the top row (Set A). Both you and the hustler now know that you have either the ace, 6, or 8.

The con artist now draws a card from the bottom row (Set C). If the hustler draws a 9, they have 3 chances in 3 of beating your card. If they draw a 4, they have 1 chance in 3 of beating your card. If they draw a 2, they have 1 chance in 3 of beating your card. Consequently, they have 5 chances in 9 of beating your card.

Therefore, no matter which row you draw a card from, the con artist has 5 chances in 9 of drawing a card that has a higher value than the card you chose. The odds at each stage favor the hustler.

The fraudster simply follows the following pattern: if you draw a card from Set B, they draw a card from Set A. If you draw a card from Set C the con artist draws a card from Set B. If you draw a card from Set A the hustler draws a card from Set C.

This beautiful proposition bet also works in a similar manner if the *lowest value* card wins. The winning cycle then is the opposite of the previous cycle: If the sucker chooses Set B, the hustler chooses Set C. If the sucker chooses Set C the fraudster chooses Set A. If the mark selects Set A, the con artist chooses Set B. By following this pattern, the hustler has 5 chances in 9 of picking the card that has the *lowest value*.

In either version of the bet, the con artist has 5 chances in 9 of winning, provided they play to the patterns given here.

Consequently, the odds are 5 to 4 in favor of the hustler winning the bet.

PROPOSITION OF THROWING 1 PAIR
WHEN ROLLING 4 DICE

"If you wait to do everything until you're sure it's right, you'll probably never do much of anything."—*Win Borden*

PROBLEM

You may be in a tavern one evening when you see a stranger all alone at the bar. You become intrigued when you see them take 4 dice (each 6-sided) from an accompanying bag.

You move a little closer to the stranger and eventually summon the courage to ask them if they are playing some dice game. They will smile and answer no, they are not playing a dice game. They will remark that when they were young, they used to roll dice quite a lot and they saw patterns emerging quite frequently when "the bones were rolled" that contravened the laws of probability.

As an example of what they mean they will propose that you roll the 4 dice and they will bet that you turn up 1 pair—in other words, 2 dice will roll to the same number.

The hustler may explain that this is unlikely to happen, because obviously it is more likely that the 4 dice will fall so that no 2 numbers on the top faces will be the same. But they will add that somehow, they feel instinctively that a pair *may* very well turn up on this occasion. So strong are their instincts that they are willing to bet $5 at even money odds that a pair will be rolled.

Should you accept the wager from this stranger?

SOLUTION

No, do not accept this bet. The odds favor the hustler.

Let's figure out the probability on this wager.

If a pair is rolled, the other 2 dice must show 2 of the other 5 values that can be rolled. This can be done in $5C2$ or 10 ways. These 2 values (often referred to by gamblers as *singletons*) can be arranged in 2 ways.

Now consider the pair that is rolled. The pair can be any 1 of 6 numbers, from 1 to 6. The pair can be arranged in $4C2$ or 6 ways.

Therefore, the total number of ways a pair can be rolled is $10 \times 2 \times 6 \times 6$, which equals 720.

The 4 dice can be rolled in 6^4 or 1,296 ways.

Consequently, the probability of rolling a pair is $720/1296$. This fraction simplifies to $5/9$. In percentage terms, that fraction equals 55.5555+ percent.

Thus, there are 5 chances in 9 that a pair *will* be rolled when rolling 4 dice. Therefore, there are 4 chances in 9 that a pair will *not* be rolled.

Consequently, the odds are 5 to 4 in favor of the hustler that a pair will be rolled when rolling 4 dice.

The odds favor the hustler.

45

CHOOSING 2 ACES OR 2 KINGS FROM 4 ACES AND 4 KINGS

"A dollar won is twice as sweet as a dollar earned."—*Paul Newman*

PROBLEM

You may be in a bar one evening enjoying a quiet drink with friends. A stranger will walk up to you and take 4 kings and 4 aces from their pocket. They will ask you to examine the cards to check that they are normal playing cards. They will then ask you to shuffle the 8 cards and to place them facedown on a table in front of you.

The stranger will then ask you to choose any 2 cards from the 8 on the table. They will bet you $10 at even money odds that you will not choose either 2 aces or 2 kings.

The stranger will probably attempt to lure you into the bet by saying something along the following lines: "You know, the odds in this bet really favor you. There are only 8 cards facedown on the table. All you must do is choose 2 aces *or* 2 kings. Simple as that! When you turn over the first card, if it is an ace, you are on your way to victory. But if it is a king, you are also on the road to success. Thus, you have 2 chances to win the game! Thus, the game really favors you, my friend."

The question is: should you agree to this bet?

SOLUTION

No way!

The number of ways of choosing 2 cards from the 8 on the table is 8C2, which equals 28.

Consider now the 4 aces. The number of ways of choosing 2 aces from the 4 aces is 4C2, which equals 6. Similarly, we find that there are only 6 ways of choosing 2 kings from the 4 kings. Thus, there are 12 ways of choosing 2 cards so that their ranks are similar.

Therefore, there must be 28 – 12 or 16 ways of choosing 2 cards where the 2 cards are of different ranks. Consequently, the odds are 16 to 12, which simplify to 4 to 3, that you will pick 2 cards that are of different ranks.

In other words, in the long term, the stranger or con artist can expect to win this bet 4 times in every 7 games.

Thus, in 100 bets, the con artist can expect to win about 57 times and lose 43. Therefore, at $10 a round, the hustler will rake in 57 $10, or $570. They will pay out $10 a round for the 43 times they lose. Thus, they lose $430. That still leaves the fraudster with a profit of $140. That equals an *average* profit of $1.40 per game.

The bet is a handy little earner for the con artist.

46

PROPOSAL THAT OUT OF 6 COIN TOSSES THERE WILL NOT BE 3 HEADS AND 3 TAILS

"The gambling known as business looks with austere disfavor upon the business known as gambling."—*Ambrose Bierce*

PROBLEM

A stranger in a bar may ask you to take a fair coin from your pocket and toss it 6 times. They will bet you $5 at odds of 7 to 5 in your favor that you will not get 3 heads and 3 tails.

They attempt to lure you into accepting the bet by stating that the odds are *even* that 3 heads and 3 tails will be the outcome of the 6 tosses. That stated outcome seems plausible. The stranger (or con artist) will say that since a coin has 2 sides, it's an even chance that a coin will fall one way or the other. Therefore, they will say it is an even chance that when the 6 coins are tossed 3 coins will fall heads and 3 coins will fall tails. The con artist will spin you some story that they are giving you 7 to 5 odds in your favor that you will *not* get 3 heads and 3 tails because they recently had a big win on the horses, and they are feeling generous.

You are listening to all of this and wondering what you should do. It seems to you that it might be worth betting on this proposition, as the odds the con artist is offering appear generous.

The question is this: Should you accept the bet?

SOLUTION

No, do not accept the bet. The odds are stacked against you and are in favor of the hustler.

Each time the coin is tossed there are 2 possibilities: it will fall heads or tails. Therefore, in 6 tosses there are 2^6 or 64 possibilities.

Here is the breakdown of the possibilities in the 6 tosses:

6 heads	1 way
5 heads 1 tail	6 ways
4 heads 2 tails	15 ways
3 heads 3 tails	20 ways
2 heads 4 tails	15 ways
1 heads 5 tails	6 ways
0 heads 6 tails	1 way
Total number of outcomes	64

We see that there are 20 ways of getting 3 heads and 3 tails.

Therefore, the probability of getting 3 heads and 3 tails is $^{20}/_{64}$, which simplifies to $^5/_{16}$. Therefore, there are 5 chances in 16 that you will get 3 heads and 3 tails.

Consequently, there are 11 chances in 16 that you will *not* get 3 heads and 3 tails. Thus, the odds are 11 to 5 against you that you will *not* get 3 heads and 3 tails.

The con artist can consequently afford to give you odds of 7 to 5 in your favor in this bet.

Suppose the stranger offers this bet 64 times at $5 a bet. The mark will win only 20 of these bets. In these 20 cases, the con artist pays out 20 × $7, or $140. But in the other 44 bets, the hustler wins. They then collect 44 × $5, or $220. Thus, their profit is a cool $80 for the 64 games. Thus, their *average* profit per bet is $^{80}/_{64}$ or $1.25.

Stay away from this bet if you want to hold on to your hard-earned cash!

47

PROPOSAL TO NOT ROLL A 7 IN 6 CONSECUTIVE ROLLS OF 2 DICE BEFORE ROLLING A 12

"The best throw of the dice is to throw them away."—*unknown*

PROBLEM*

You may be in a bar some evening and a stranger may enter. Unknown to you and the other patrons of the bar, this stranger is a con artist. They are there looking for business. In other words, they are looking for an easy way of collecting some greenbacks for themselves. They will probably eventually engineer the conversation on to the subject of gambling.

Eventually the stranger may say something along the following lines: "If one rolls 2 fair, 6-sided dice the probability of getting a total of 7 on the 2 dice is 6 chances in 36. The probability of rolling two 6s (a total of 12) is just 1 chance in 36. (Both of these probabilities are correct.) Therefore, the probability of rolling a 7 is 6 times greater than rolling a 12. Consequently, the probability of rolling 7 with 2 dice 6 consecutive times is equal to the probability of rolling 12 just once."

The con artist may then offer you the following proposition: they will bet you $10 at even money odds that you will *not* roll 7 with 2 dice 6 consecutive times before you roll a 12.

Should you accept the bet?

---

*Source: "Probability—Dice," Wizard of Odds, https://wizardofodds.com/ask-the-wizard/ probability/dice/, accessed April 2, 2017.

SOLUTION

No way!

The probability of rolling a 7 or 12 (and ignoring all other results) is $6/7$. Why? Well, there are 6 ways to get a total of 7 rolling 2 dice and only 1 way to get a total of 12. Thus, the probability of rolling 7 is 6 times greater than the 1 chance of rolling 12. Therefore, of the 7 outcomes, 6 of them involve rolling 7, and 1 involves rolling 12. Thus, the probability of rolling 7 before 12 is $6/7$.

The probability of rolling 6 consecutive 7s is therefore $(6/7)^6$, or $46656/117649$. (This fraction equals 0.3965+.)

Therefore, the probability of *not* rolling 6 consecutive 7s, but of rolling one 12 in those 6 rolls, is 1 (*certainty*) – 0.3965+. This equals 0.6034+. In percentage terms this equals 60.34+ percent.

In other words, the probability is 60.34+ percent that you will *not* roll 6 consecutive 7s before you roll 12. The probability that you will roll 6 consecutive 7s before you roll 12 is 39.65+ percent.

This means that in the long run the hustler will win about 60 of every 100 bets they make and lose about 40. Thus, the odds are about 60 to 40, or 3 to 2, in favor of the con artist.

If the con artist proposes this wager 100 times, at $10 a bet, at odds of even money, the hustler can expect to profit by about $200.

48

PROPOSAL OF TOSSING A FAIR COIN 4 TIMES WITHOUT GETTING 2 HEADS

"It's choice—not chance—that determines your destiny."—Jean Nidetch

PROBLEM

You may be at a social gathering one day when a friendly stranger walks up to you and chats to you about various topics. Eventually they will bring the subject around to the theory of probability.

They may tell you that when they were younger, they studied probability at some college or other. Ever since those happy bygone days, when the bills came in someone else's name, they say that they often perform probability experiments to confirm some unproven theory about "the law of averages" that they learned from an eccentric professor when they were just 18 years old. They go on to state that they would like to do a simple experiment in probability now, to see if the theory still holds, and asks if you would care to volunteer in the exercise.

Intrigued by what you hear you agree to do so.

The stranger asks you to produce a fair coin. They then ask you to toss the coin 4 consecutive times and each time to note whether the coin falls heads or tails.

They will probably add something along the following lines: "Your coin is obviously a fair coin. If we toss it 4 times in the air, the odds we will get 2 heads and 2 tails are obviously even. However, I am prepared to bet you $10, at odds of even money, that if you toss your coin 4 times you will *not* get 2 heads. We are *not* talking about consecutive heads here. I am betting that you will *not* get 2 heads *at all* in the 4 tosses. Period. It is as simple as that! If you get 2 heads, I will give you 10 bucks. If you fail to get 2 heads, you give me $10. That is fair in my book. Let's play the game 3 times at $10 a round. You might pick up a handy 30 bucks, if lady luck is smiling on you today."

You are sitting there wondering that perhaps this is your lucky night.

This question sprinting through your mind is: Should I accept the bet?

SOLUTION

No way! The odds quoted by the con artist are incorrect.

Each time a genuine coin is tossed there are 2 possible outcomes. Thus, if the coin is tossed 4 times the number of possible outcomes is 2^4, which equals 16.

Letting H represent heads and T represent tails, here are the 16 possible outcomes to the 4 tosses of the coin: HHHH, HHHT, HHTH, HTHH, THHH, HHTT, HTHT, TTHH, THTH, THHT, HTTH, TTTH, TTHT, THTT, HTTT, and TTTT.

You will find that there is 1 way that results in all heads. There are 4 ways that result in 3 heads and 1 tail. There are 6 ways that 2 heads and 2 tails can fall. There are 4 ways that result in 1 heads and 3 tails. There is 1 way that no coins fall heads; in other words, all 4 coins fall tails.

The total number of possible outcomes is 16. These are the only possible outcomes when a fair coin is tossed 4 times. In just 6 of these ways do we see 2 heads and 2 tails falling. Therefore, the probability of you obtaining 2 heads and 2 tails is $\frac{6}{16}$, which simplifies to $\frac{3}{8}$.

Consequently, the probability that you will *not* get 2 heads and 2 tails is $\frac{5}{8}$. This equals 0.625. In percentage terms, this equals 62.5 percent.

Saying the same thing another way, the probability that you will *not* get 2 heads and 2 tails is 62.5 percent. The probability that you will get 2 heads and 2 tails is 37.5 percent.

The odds are about 62 to 38, which simplify to 31 to 19, that the swindler will win the bet.

For every 100 times the con artist offers this little wager, they can expect to win the bet about 62 times and lose about 38 times. At $10 a bet, they win about $620 and lose about $380, yielding a tidy profit of $240.

Thus, at $10 a round, the *average* profit per bet is about $2.40.

As the adage states, "Nice money if you can get it!"

49

THE SWINDLE KNOWN AS THE GAME OF 31

"Lest men suspect your tale untrue, keep probability in view."—John Gay

PROBLEM

You may be enjoying a quiet drink in a bar one evening when a stranger walks in. They will engage you in conversation and eventually bring the subject around to gambling. They will tell you that they have loved gambling since their childhood days and that sometimes they will even bet on even money propositions. As an example of their preparedness to gamble on almost anything they will take a deck of cards from their pocket and lay out the following 24 cards faceup: Ace, ace, ace, ace, 2, 2, 2, 2, 3, 3, 3, 3, 4, 4, 4, 4, 5, 5, 5, 5, 6, 6, 6, 6.*

The stranger (or con artist) will now ask a customer in the bar to play a Game of 31, just for fun. The hustler will point out that no money at all will be involved. They will tell you that the game is played as follows: You first take a card, then they take a card, then you take again, then the con artist. And so on. Each time a card is taken there is a running total kept. Thus, if your first card is 3, the con artist may take a 2, and say "5." If you then take a 6, you say "11." The con artist takes an ace and counts "12." And so on. The first player to count to 31 is the winner.

Suppose a customer agrees to play for fun, and the operator of the game has the first move. The operator first takes a 3 and the customer takes a 2, counting "5." The operator takes a 5 and counts "10." The customer takes a 3, counting "13." The operator takes a 4 and counts "17." The customer takes a 5 and counts "22." The operator takes a 2 and counts "24." The customer takes a 4 and counts "28." The operator takes a 3 and counts "31." Thus, the operator wins the game.

The operator plays another couple of games for fun against other bar customers and wins each game.

If you have been watching each game closely you may have noticed that each time the operator won a game, they seemed to have been on a running total that equaled 3, 10, 17, and 24 during the game.

You soon catch on to the game. You realize that 3, 10, 17, and 24 are key numbers. The player that can get their running total to equal these numbers and "stay on course" during a game is certain to win. For example, if player A is on 24, there is no way that player B can score 31, because the largest value any card has is 6. Therefore, whatever card B chooses, it is a simple matter for A to choose another card and score 31.

---

*Source: Percy Abbott, *Abbott's Magic for Magicians: Occidental and Oriental Mysteries* (Colon, MI: Abbott's Magic Novelty, 1934).

With this insight into the Game of 31, you have gained some confidence.

You now feel increasingly sure that if you can play first, you can beat the operator at their own game. Furthermore, if you can get the operator to play for big bucks you could make a "killing" in financial terms.

Now that you know the rule, you watch the operator play a few more games. Each time the operator plays in such a manner that they try to achieve a running total of 10, 17, and 24, confirming your belief about the strategy to adopt to win the game. Eventually, of course, the operator gets onto their key numbers and is the first to reach 31 and, therefore, wins the game.

At this point the operator begins to brag about their prowess at this game and says loudly that they can beat any opponent, playing for fun *or* for any sum of money that anyone cares to gamble with.

You now believe, of course, that if you were to play the operator and you had the first move, then you would win this game by playing to the key numbers each time it is your move.

It seems a big opportunity has come your way. You stand a good chance to make a large bundle of dollars.

You state that you are prepared to challenge the operator if you are allowed the first move.

The question is this: Should you play this game for money against the operator?

SOLUTION

No, do not play this game with the stranger!

You have watched the operator play this game and you have been sufficiently clever to spot the mechanics of the game in order to beat the operator, provided you get the first move.

Let us assume you volunteer to play the operator for fun first, or for a few dollars, provided you get the first move. The operator agrees to your condition and plays a game with you, which you win.

The operator then pretends to be flabbergasted at the result. They state loudly that your win was just a *fluke* and they feel sure that should you play again for "real money" that they would easily win. They offer to play the game for $50 at even money odds.

In this way—unknown to you, or the others at the bar—the operator, who is really a con artist, has set the bait.

You fall for the trap and agree to play the Game of 31 for $50 if, and only if, you have the first move.

The hustler appears to reluctantly agree to this condition.

The swindler places $50 on a nearby table and asks you to match it. You confidently place your stake of $50 down on top of the pile of dollars the operator has placed.

You take the first card, which is a 3. The operator turns a 4 counting "7." You take a 3 again, counting "10," and the operator turns 4, and counts "14." You take a 3 counting "17." The operator takes a 4 counting "21." You take a 3 again, counting "24." At this point you are feeling ecstatic, as you can see almost nothing but a fistful of dollars coming your way. The operator takes another 4 and counts "28."

You reach for a 3 to win the game but find that there are none! All the 3s have been used! It suddenly dawns on you that no matter what card you take you cannot reach 31. In other words, your situation is hopeless. You cannot win!

All you can do is take either an ace or a 2. If you take an ace, the operator places a 2 on top of it. If you take a 2, the operator places an ace on top of it. In either case the operator wins the game and collects the $100 on the table.

The horrible thought crosses your mind that you have been conned. The "operator" is nothing but a ruthless, cunning con artist who makes their living from swindling honest folk like you!

If you see the Game of 31 being played, do not get involved, if you want to hold on to your dollars.

50

PROPOSITION THAT TAILS, HEADS, HEADS WILL APPEAR BEFORE HEADS, HEADS, TAILS

"All nature is but art, unknown to thee; All chance, direction, which thou cannot see."—*Alexander Pope*

PROBLEM

You may be in a bar one evening when a stranger walks in and engages you in conversation. They will tell you they are a bettor and will bet on most things in life, even events that are unlikely to happen. As an example, the stranger offers the following proposition bet to you: They will ask you to take a coin from your pocket and to repeatedly toss the coin, noting the results of the first, second, and third tosses; then the results of the second, third, and fourth tosses; then the results of the third, fourth, and fifth tosses. And so on.*

Letting H represent *heads* and T represent *tails*, the hustler will tell you that you might get the following results if you were to make 12 tosses of a fair coin: H, T, H, T, T, H, T, H, H, T, T, and H.

Having patiently explained what they want you to note, the con artist now bets $10 at even odds that the sequence THH will appear *before* HHT.

Being mathematically minded, you will probably work out reasonably quickly that you are dealing with 8 possible triples here: HHH, HHT, HTH, HTT, THH, THT, TTH, and TTT.

You quickly realize that the bet is basically the con artist stating that the symbols THH will appear before HHT.

Because you are a mathematician (or at least a person interested in mathematics), you are inclined to think that the appearance of 1 sequence of 3 symbols is as likely as any other. You also happen to like a bet now and again. Consequently, you may find that you are inclined to accept the bet.

The question is this: Should you accept the bet?

---

*Source: Ross Honsberger, *Mathematical Plums* (Washington, DC: Mathematical Association of America, 1979), chapter 5: Some Surprises in Probability.

SOLUTION

No way!

This is one of the finest sucker bets you will ever encounter.

Believe it or not, it is 3 times more likely that the sequence THH will appear before HHT. The odds are therefore 3 to 1 in favor of the con artist!

How are these unexpected probabilities calculated?

It is not difficult to work the whole thing out.

Consider the first 2 tosses of the coin. The result of these 2 tosses must either be HH, HT, TH, or TT. These are the only 4 possible outcomes to the first 2 tosses.

Suppose the first 2 tosses produce the sequence HH. For the con artist to win, the sequence HH must be preceded by a T. But that is impossible, because the first 2 tosses were *heads*, producing HH.

Now if the third toss results in a *tail*, you (and not the con artist) will win the bet. If the third toss is H and the fourth toss is T you will win the bet. If the fourth toss is H and the fifth toss is T you win the bet. Sooner or later the sequence HHH . . . will be followed by a T and therefore you will win the bet. Summing this up, if the first 2, 3, 4, 5, . . . tosses produce *heads*, you will win the bet, because eventually . . . a *tail* will be tossed, and thus the sequence HHT will emerge.

However, suppose the first 2 tosses result in HT, TH, or TT.

You continue tossing the coin a third time, a fourth time, a fifth time, a sixth time, and so on. Suppose somewhere in this series of tosses, say the tenth, eleventh, and twelfth tosses, the sequence HHT appears. Consider now the ninth toss. If that is a T, we have the sequence THH appearing *before* the sequence HHT, and therefore the con artist wins the bet. However, if the ninth toss is H, we have the sequence HHH beginning with the ninth toss. Consider now the eighth toss. If that is T then the sequence THH has appeared before the sequence HHT, and the hustler wins the bet.

The only way you win the bet is if the sequence HHH, when it first appears, was preceded by H right back to the first toss.

But we have just stated that the first 2 tosses were HT, TH, or TT. Therefore, it is impossible that the sequence HHH is preceded by H all the way back to the first toss.

Therefore, when the first 2 tosses result in HT, TH, or TT, the con artist wins the bet. Starting with these 2 outcomes, we see that there are 3 different ways the hustler can win the bet.

Only in the 1 case where the first 2 tosses result in HH do *you* win the bet. Therefore, the odds are 3 to 1 that the hustler will win the bet.

It is a fantastic sucker bet, for the con artist!

Don't get taken by it!

51

BRIEF HISTORY OF PROBABILITY THEORY AND SOME EXAMPLE SCAMS

The theory of probability is extremely important and useful in the modern world. Medicine, engineering, economics, and social planning are but a few of the many areas where probability theory is paramount.

Thus, probability theory is an essential component to the thinking person's understanding of how the world operates.

However, of the many branches of mathematics, the theory of probability is also known to mathematicians as being the one branch of that most beautiful of subjects where even professional mathematicians can slip up.

The origins of probability theory go back to France in the seventeenth century.

In that century, Blaise Pascal (1623–1662) a French mathematician and physicist, was asked by a gambler about the probabilities of obtaining specific results in throwing dice. The subsequent work that Pascal did on the problem was the beginnings of what has become known throughout the world today as the *theory of probability*.

The gambling problem was posed to Pascal by Antoine Gombaud, Chevalier de Méré.[1] He was a French nobleman with an interest in gambling questions. Gombaud drew Pascal's attention to an apparent contradiction in gambling folklore at the time concerning a dice game. The dice game involved throwing a pair of dice 24 times. At the time, gamblers were aware of a well-established gambling rule that stated that betting at even money odds on a double 6 (2 dice falling with the number 6 uppermost on each die) in 24 throws of 2 dice was profitable. However, Gombaud told Pascal that his calculations led him to believe that the old gambling rule was incorrect. Gombaud asked Pascal to calculate the true odds of this bet.

In response Pascal wrote to an amateur, but very able, French mathematician, Pierre de Fermat (1607–1665) concerning the problem. (Pierre de Fermat's name later became famous for a mathematical problem known as *Fermat's Last Theorem*, which was not solved until 1994. In that year Andrew Wiles and his former student Richard Taylor solved the problem that had been unsolved for more than 3 centuries.) The correspondence between Pascal and Fermat led to the formulation of the basic principles of probability theory. Prior to that correspondence, Italian mathematicians had done little work on games of chance. It was not until Pascal and Fermat got involved that the theory of probability as we know it today came into being.

The wager that one will obtain at least 1 double 6 in 24 throws of a pair of dice is hardly ever used today as a proposition bet. The reasons for this are varied but reduce to 2 main factors. First, it takes too long to roll a pair of dice 24 times. Plus, the odds of winning the bet are insufficiently favorable to the hustler to justify offering the wager to a mark.

The calculation of the probability of obtaining a double 6 in 24 throws of 2 dice proceeds as follows: The probability of *not* getting a double 6 in 1 throw of a pair of dice is $^{35}\!/_{36}$. Therefore, in 24 throws the probability is $(^{35}\!/_{36})^{24}$. This equals 0.50859+. This is the probability that there will *not* be a double 6 in 24 throws of 2 dice. Thus, to obtain the probability that there will be a double 6 in 24 throws, we subtract 0.50859+ from 1 (*certainty*). The answer is 0.49140. This is less than 0.5. Therefore, it is slightly less than 50 percent likely that one will throw a double 6 when rolling a pair of dice 24 times.

If we perform a similar calculation for 25 throws of a pair of dice, we find that the probability of throwing a double 6 is 0.50553+. Thus, it is very slightly more likely that 1 will throw a double 6 with 25 throws of a pair of dice. The very slim advantage, however, makes this wager unsuitable for a proposition bet.

Bets with dice are, however, used widely by con artists to deceive marks in the world today. The following wager with 1 die is an example of this. Con artists have used the bet for profitable purposes for generations. Here is how the bet goes: The grifter asks the mark to select any 6-sided die of their choice. The con artist then tells the mark that in a moment they are going to ask them to roll this 1 die several times to eventually turn up all 6 numbers on the top face. The swindler tells the mark that the chances of getting any 1 number on the top face is obviously 1 chance in 6. But the scammer points out to the mark that they (the scammer) are in a generous mood and are giving them, not 6, but 12 rolls of the die to roll all 6 numbers. If the mark rolls all 6 numbers on the die, they win the bet. But if they don't roll all 6 numbers, they lose the bet. The con artist bets the sucker $5 at even money odds that they (the sucker) will not roll all 6 numbers in 12 rolls.

The mark will probably reason that since there are 6 faces on the die, it seems reasonable to assume that they will eventually roll every number on the die given that they have 12 rolls to do so. The mark is therefore likely to accept the bet.

However, surprisingly, it takes—on *average*—about 15 rolls of a die to roll every number from 1 to 6.

When the mark rolls the die the first time, they are certain to roll a number on the die. On the second roll, the probability they will roll a different number from their first roll is $^5\!/_6$. Therefore, when the mark rolls the die a second time, they must—*on average*—roll the die $^6\!/_5$ times to obtain a number different from the first number they rolled. On the third roll they will on average have to roll the die $^6\!/_4$ times to obtain a number different from the first 2 rolls. And so on.

Therefore, the number of rolls they can be expected to make before rolling all 6 numbers on the die is:

$$\frac{1}{1} + \frac{6}{5} + \frac{6}{4} + \frac{6}{3} + \frac{6}{2} + \frac{6}{1} = 14.7$$

The mark therefore needs about 14.7 or 15 rolls on *average* in order to roll all 6 numbers on 1 die. With 15 rolls of a die the probability that all 6 numbers will eventually show is 64.4212+ percent.[2] Thus, *if* the proposition bet involved an invitation to the mark of rolling the die 15 times, the odds of rolling all 6 numbers on the die would favor the mark winning, by approximately 16 to 9. Thus, in 25 such bets, the mark will win about 16 of them and lose 9. Consequently, no self-respecting professional dice hustler will offer a proposition bet with these terms.

Generally, the mark is given a bet that they must roll all 6 numbers on a die at least once in 12 rolls of the die. The mark is on a losing ticket here, as they need, on *average*, 15 rolls to win the bet.

Calculating the exact probability that the mark will have rolled all 6 numbers of a die in 12 rolls is difficult. The exact probability can be achieved by using a technique in advanced mathematics called *generating functions*. By using this technique, it is found that the probability of a mark rolling all 6 numbers in 12 rolls of a fair die is 0.43781568+.[3]

In other words, the mark has a 43.78+ percent probability of success with this bet. Of course, this means the con artist has a 56.22+ percent probability of winning the bet. The probability of winning favors the con artist! On *average* the con artist will win 56 of every 100 such bets.

Proposition bets with 2 dice are popular with hustlers. One of the oldest cons goes as follows: The con artist explains to the mark that a 7 can be obtained in the following 6 ways by rolling 2 dice: 1, 6; 2, 5; 3, 4; 4, 3; 5, 2; 6, 1. The hustler tells the mark that they are offering them 3 rolls of the 2 dice. The grifter explains that the 2 dice must fall in 1 of 36 ways, but a 7 can be obtained in 6 ways. The con artist goes on to say that if the 2 dice were rolled just once the mark would have 6 chances in 36 of winning. But the mark is going to get 3 rolls. This gives them 18 chances, so the probability of the mark winning is even. The con artist then bets the mark $5 at even money odds that they will not roll a 7 with 2 dice in 3 rolls of the dice.

The chances of the mark *not* rolling a 7 in the 3 rolls is $(^{30}/_{36})^3$, which equals 0.5787+. Thus, the probability of them *not* rolling a 7 on 3 rolls of 2 dice is 57.87 percent. Therefore, the probability the mark will roll a 7 on 3 rolls of 2 dice is 42.13 percent. The odds in this bet are about 4 to 3 in favor of the hustler.

The following is a beautiful and easy-to-understand con with 2 dice. The swindle goes as follows: the con artist will produce 2 dice and ask the mark to roll both of them. The hustler will charge the mark $5 for each roll of the 2 dice.[4]

The swindler tells the mark that they can bet that the 2 dice will—when rolled—show a total sum of 2, 3, 4, 5, or 6. If the mark makes that bet, and then happens to roll a total sum of 2, 3, 4, 5, or 6, they win, and the con artist will pay them at even money odds. In other words, the mark will win $5 and get back their $5 stake.

The hustler will then tell the mark that if they wish they can bet that when they roll the 2 dice they will roll a total sum of 8, 9, 10, 11, or 12. If they agree to that bet and they succeed in rolling the 2 dice so that a total sum of 8, 9, 10, 11, or 12 show, they also win, and the grifter will pay them at even money odds. In other words, the mark will win $5 and get back their $5 stake.

The con artist then tells the mark that they can also make the following bet, if they so wish: The mark can bet that the 2 dice—when rolled—will show a total sum of 7. If the mark makes this bet and they subsequently roll a 7, then the mark wins. In this case the hustler will pay them at odds of 4 to 1. In other words, for a $5 stake, if the mark wins this form of the bet, they will receive $20, plus their $5 stake back.

The hustler tells the mark that they really should consider making one of these bets, as the odds are in their favor. The con artist will probably add that there is hardly any way the mark can lose in this bet.

Having considered the matter for some time, there is a strong likelihood that the mark will accept one of the above bets. If the mark does agree to one of these wagers, they are onto a losing ticket.

To see why, consider the following: if one rolls 2 dice, there are a total of 15 ways of rolling a total sum of 2, 3, 4, 5, or 6. But there are 36 ways the 2 dice can fall. Therefore, the probability that the mark will win is 15 chances in 36. Thus, the odds of their winning are 21 to 15, which simplify to 7 to 5 against the mark. But they are only being paid at even money odds. Performing this bet regularly, the hustler will win 7 bets in every 12. At $5 a bet, over the long term, the fraudster will win $35 and lose $25 in those 12 bets. That is a profit of $10.

The same principle applies to the proposition bet that with 2 dice, the mark will roll a total sum of 8, 9, 10, 11, or 12. There are just 15 ways that they can do this. Therefore, the chances of the mark rolling a total sum of 8, 9, 10, 11, or 12 are 15 in 36, which simplify to 5 in 12. Once again, the odds of the mark winning are 7 to 5 against them. But they are only being paid at even money odds. As with the previous bet, the con artist will over time make a profit of $10 for every 12 bets at $5 a bet.

What if the mark bets that when they roll the 2 dice, they will roll a total sum of 7? The mark may very well accept this bet as they are being offered odds of 4 to 1 in their favor. The number of ways a 7 can be rolled with 2 dice is 6. Therefore, the probability they will succeed in doing this is 1 chance in 6. Consequently, the *true* odds of the mark doing this are 5 to 1 against. But they are only being paid at odds of 4 to 1 in their favor. Thus, over the long term, for every 6 bets, the hustler will win 5. At $5 a bet, this will mean the fraudster will rake in $25, plus their stake of $5. The 1 bet in 6 that they lose, they will pay out 4 times $5, or $20, plus the mark's stake of $5. Thus, for every 6 bets at $5 a bet, the con artist can expect to make a profit of $5. That equates to an average profit of about 0.83 cents per game for the hustler. If the hustler is performing these bets regularly, they are onto a winning ticket.

Of course, con artists have used other nonmathematical but clever methods to cheat their marks of their money.

In past centuries, a con artist would often perform the cups and ball routine in a public street. The scam is regularly performed on the streets of London, in the United Kingdom, mainly cheating tourists out of their hard-earned cash. One sometimes comes across magicians performing this con purely for entertainment purposes in their magic shows. At these shows the magician will perform the scam and, in accordance with the universal magicians' code, will not reveal how the scam works. In such shows, the magician usually warns their audience not to be enticed into playing this rigged gambling game.

The cups and balls routine runs as follows: in full view of the mark, the fraudster will hide 1 ball under 1 of 3 cups, which are situated on a small fold-up table (which can be folded up quickly should someone call the police). The hustler will then quickly slide the cups around for 20 seconds or so, without lifting them. The mark's job is to follow the cup that covers the small ball. The con artist will then bet the mark, say, $5 at even money odds, that they cannot point to the cup that covers the concealed ball. Of course, no matter what cup the mark points to, they will find they are invariably wrong.

When the mark loses, they are tempted by the con artist to play again to win their money back. If the mark agrees to this, they will lose a second time. This may go on for a few times more. Eventually the con artist will raise the stakes in order to tempt the mark to continue playing to give the mark a chance to recover their losses. But in agreeing to this, the mark is doomed. The hustlers who perform this swindle are usually so good that it is near impossible for the mark to win.

The reader is advised not to play this game on a public street. If you do, you are almost certain to lose your cash. The scam is illegal in most countries, but that does not stop the scammers from defrauding honest members of the public.

A similar scam is the walnut shells and pea con. In this fraud, the scammer has 3 walnut shells placed on top of a small fold-up table. One of the walnut shells covers a tiny pea. In full view of the mark the hustler places 1 pea under 1 of the shells and then quickly slides the shells around for a few moments. The mark's task is to keep track of the walnut shell covering the pea. When the hustler finishes sliding the walnut shells around, they will bet the mark $5 at even money odds that they cannot point to the walnut shell covering the pea. The mark will find that no matter which walnut shell they point to, they cannot pick the one covering the pea. Once again, once the scammer gets the mark involved in the game, the fraudster gradually raises the stakes. If the mark continues to play, their losses dramatically increase.

A version of the scam is performed with bottle caps in New York City, as well as other locations in the United States that are popular with tourists. This version of the scam is also popular in many European cities. There have even been versions of the con performed with matchboxes. Tourists are advised not to get involved in these games if they want to hold on to their cash.

These scams go back hundreds of years. They were originally perpetrated by scammers at fairs and race meetings and anywhere large crowds were gathered. But today these fraudsters are just as much at home on the public streets of large cities. The scams are presented as genuine gambling games, where the marks apparently have a genuine chance of winning. But these games are not gambling games; they are scams. The operators of these games are thieves. They have no qualms about taking your money. Protect yourself by staying away from them and their scams.

Another one of these cons involves what is known as the *Endless Chain*. The con artist produces a metal chain connected at both ends, which is about a meter (3.28 feet) in length. The con artist places the chain down on a little table and makes a figure of 8 loops. They demonstrate what they want the mark to do by saying, "All you need do is pick 1 of those 2 loops that I make with this chain and place your index finger in the loop. I will then pull at the other end of the chain. If your finger gets caught in the loop, you win $5. If, however, you pick the loop where your finger does *not* get caught in the loop, you lose and you pay me $5. Now before we begin, we both will place our $5 bills here on the table. The winner takes the 2 $5 bills, of course."

The mark will find that no matter which loop they pick, they will always pick the loop in which their finger does *not* get caught. The mark will thus constantly lose the bet. The operator of this scam is a fraudster. Do not get involved with this hustle.

One of the most famous scams on the streets is that known as *Three Card Monte*. The con artist has 3 cards, usually 2 red queens and an ace, on a small fold-up table—usually in a public street. The trickster shows the 3 cards to the mark. The fraudster then throws the cards facedown onto the small table. They bet the mark a sum of money that they (the mark) cannot point to the ace. No matter how they try, the mark fails to locate the ace.

If the mark (by accident or good luck) points to the correct card, an onlooker in the crowd (this supposed onlooker is known in the gambling world as a *shill*), on seeing this, will bet a higher amount on another card. Unknown to the mark, the shill is secretly working in tandem with the hustler. The con artist operating the trick will then of course accept the higher bet and turn over the card that the shill pointed to. This card, of course, proves to be the incorrect card. But by operating this procedure, the operator effectively stops the turning over of the correct card pointed to by the mark. The mark sees all this happening in front of their eyes, but of course, they are unaware that they are being conned and therefore suspect nothing. The mark will therefore never know if the card that they pointed to was the correct card. This is all done to keep the workings of the trick as secretive as possible.

Occasionally, over the years, magic suppliers have released booklets explaining the various scams perpetrated against the public on public streets. The booklets were designed to teach magicians how to perpetrate these scams, so that they might use the scams as a part of their magic shows. In 1983 a magic company in Dover, England, named Supreme Magic Company (unfortunately the company is no longer in business) issued a booklet titled *The Endless Chain*. It was one booklet among many in the Supreme Know-How Series. The company also issued an educational booklet around that time titled *Three Card Monte*.

There are numerous other scams that are regularly perpetrated against members of the public. Here is one of the best.

An old, shabbily dressed person enters a bar or restaurant carrying an old fiddle.[5] It appears obvious to the customers in the bar (including the bartender) that the elderly person is a street musician. They proceed to sit down at a table and order a meal. Having eaten, they then call the server and say that they have left their wallet at home. They also tell the server that fortunately they live nearby and will go and get their wallet and return with it to pay for their meal. They offer their only worldly possession—the old violin that provides them with their livelihood—as collateral. The server believes the stranger's story and agrees to take custody of the old violin. The stranger leaves.

Shortly after, another stranger enters the restaurant. They see the old violin and, pretending to be a collector and an expert on rare musical instruments, declare that the old violin is an extremely rare kind and is worth at least $10,000. This second person says they must, unfortunately, leave the restaurant as they have an important business appointment but leave their business card with the server, asking them to phone them when the owner of the violin returns as they (the musical instrument expert) are considering buying the violin and are prepared to pay a large sum of money to acquire the rare and valuable instrument. The second stranger now leaves.

The server (or mark) is still in possession of the violin, which they now believe is worth at least $10,000. When the first person returns to pay their bill, the mark offers to buy the violin from them. This person says they are reluctant to sell the instrument since they have had it for a lifetime, and it is their sole way of making a living. But eventually they say they are willing to sell the violin if they can get a decent price for it.

The server and the stranger begin to haggle over the price of the violin. Eventually the violin is sold for a very significant sum of money, but not so significant that the server can't still make a large profit (or so they believe) when they sell the violin to the second man.

The server hands over the purchasing price of the violin, perhaps $400 or $500, and the stranger accepts the money and leaves. They are never seen in the area again. The second stranger, of course, never turns up to inquire about buying the violin.

It soon becomes clear to the server that they have been conned by 2 scammers working together as a team. The result is that the 2 con artists have lost the old violin (which turns out to be almost worthless) and have pocketed a healthy profit for themselves as a result of their couple of hours of work.

The server (or mark) is, of course, left with a worthless violin that nobody wants.

This con is still perpetrated (sometimes in slightly different forms) around the world today.

Scammers sometimes perform simple magic tricks to perpetrate a fraud. Here is one example of this type of scam. The workings of this con were passed on to me by the late Martin Gardner. Martin published the workings of the scam on page 259 in his book *Mathematical Circus*, first published in the United States by Alfred A. Knopf, New York, in 1979. Apparently, the scam was quite often perpetrated in bars throughout the United States in the 1930s and 1940s. The con is so good that it would be surprising if it was not still performed somewhere in the world today.

Here is how the swindle works: Two con artists enter a bar separately. One of them sits at one end of a bar, orders a drink, and soon starts performing magic tricks to amuse the bartender and a few nearby customers.

Having done a few tricks the fraudster announces that they will now perform a truly amazing trick. The trick requires a $50 bill. The con artist asks the bartender to loan them such a bill from the cash register behind the bar. The bartender takes a $50 bill from the cash register. The con artist then asks the bartender to copy down the serial number on the $50 bill on a piece of paper and to sign their name (the bartender's name) on the bill. The con artist explains they are asking for this to be done so that when the bill is returned to the bartender a few minutes later, they (the bartender) will know it is the bill they took from the cash drawer.

When the bartender signs the $50 bill, they hand it to the con artist, who then apparently seals the bill in an envelope. The bill, however, is secretly passed through a slot in the back of the envelope and is palmed by the con artist. ("Palmed" is a word used by stage magicians to describe how an item is secretly hidden in one's hand, unknown to the audience.) The con artist then proceeds to burn the envelope in front of the bartender and fellow customers in an ashtray on the bar counter, apparently destroying the $50 bill that the bartender and others believe is in it.

As the envelope burns the fellow scammer of the con artist inconspicuously passes closely by the first scammer. As this is done, the first con artist secretly passes the $50 bill to the second scammer, who proceeds to walk up to the far end of the bar counter. The second scammer uses the $50 bill to pay another bartender for a drink that costs (at most) a couple of dollars. The second con artist collects the change from the second bartender.

Meanwhile at the other end of the bar the first con artist tells the bartender (the mark!) to go up to the far end of the bar and look into their cash register located there, where they will find the original $50 bill used in the trick.

The $50 bill is found. Its serial numbers match that which was written down earlier by the bartender and furthermore the bill is signed with the bartender's name. The bartender and many customers are, of course, astonished at the trick.

Of course, as the bartender and customers marvel at the trick, the 2 con artists quickly leave the bar with a tidy profit of about $48 for their enterprising work.

Con artists have other clever and confusing ways of deceiving the gullible public into parting with their money. One such con is known as the change-raising scam.[6] The con involves the scammer's getting back more change from a transaction than they are due.

I give here a number of examples of how this con works: A customer enters a store or a gas station with 2 $50 bills in their pocket. The grifter hands the clerk a $50 bill and asks the clerk for 10 $5 bills in exchange for the $50 bill. The clerk goes to the cash register and takes 10 $5 bills from it. They then take the $50 bill from the con artist and count out 10 $5 bills on the counter to give to the grifter. The con artist takes the 10 $5 bills. The clerk takes the $50 bill to the cash register but just before she puts the cash in the drawer the grifter says: "I've made a mistake. I have found a few more $5 bills in my pocket. I do not need all these $5 bills. Please give me back the $50 bill and I will give you these 10 $5 bills." At this point, unknown to the clerk, the fraudster secretly exchanges their second $50 bill for 1 of the $5 bills in their pocket.

The fraudster hands a bundle of $5 bills (presumably 10 of them) to the clerk and takes back the (first) $50 bill from the clerk. The clerk counts the $5 bills, and finds that there are 9 of them, plus a $50 bill, in the bundle. "You have made a mistake here," the clerk says. "You have given me $95. You should only have given me 10 $5 bills."

The con artist apologizes and says, "I have too many of these bills in my pocket. They are so cumbersome. Look, you have $95 there in your hand. Here is another $5." The hustler then hands the clerk another $5 bill saying, "Just give me $100 and we will both be right. I apologize for wasting your time."

The clerk takes the 10 $5 bills, plus the $50 bill to the cash register, takes out a $100 bill, and hands it to the con artist. The grifter leaves, having just swindled the clerk's store out of $50.

The exchange might appear confusing, but that is precisely what the grifter is hoping to do: to cause confusion in the mind of the store clerk. You see, when the grifter gives back the stack of 10 $5 bills and receives back the first $50 bill from the clerk, the swindler is at that point in time, even with the store clerk. Neither owes the other anything. But the clerk spots that the con artist hands her only 9 $5 bills and 1 $50 bill. That is a total of $95. The clerk, being honest, points this out to the grifter, saying that they paid $45 too much. The fraudster apologizes and passes 1 more $5 bill to the clerk. At this point the clerk owes the con artist the $50 extra they gave her with the $10 bills. But the grifter just tells the clerk that she has now $100 in her hands and asks for a $100 bill in exchange. Of course, when the clerk does this, the fraudster has deceived her out of $50. However, the clerk will probably not realize the fraud until later when the cash register is checked, and the con artist is long gone.

The whole swindle is achieved because the con artist has created confusion in the clerk's mind by performing a couple of different transactions at one time.

Here is another example of how the con artist defrauds a victim using the change-raising scam.

A grifter enters a store or a gas station and pays for an item that costs approximately $1 or less with a $10 bill. The con artist asks for 9 $1 bills (plus any relevant coins that may be due) as their change. The store clerk duly places the correct change and 9 $1 bills on the store counter. The fraudster then rummages in their pockets and tells the clerk they have another $1 bill, and that they will put this with the 9 $1 bills on the counter and exchange it for a $10 bill. While saying this the con artist asks the clerk for the $10 bill and then gives the clerk the 9 $1 bills that the clerk had placed on the counter, plus the $10 bill the clerk gave them. The clerk, noticing this, will assume that it is a mistake on the customer's part and will offer the $10 bill back to the customer in exchange for a $1 bill. The customer then will probably say: "Oh yes, here's another $1 bill I have found in my pocket. Look, you take this $1 bill. Added to the $9 in your hand, that makes $10, plus the $10 bill you have in your other hand, makes 20. Just give me a $20 bill in exchange, and we will both be right."

The clerk complies with the customer's request. The customer then walks from the premises with a $10 profit.

To see how the clerk was cheated, recognize that when the clerk offers to return the $10 bill in one of their hands to the customer and receives 1 additional dollar from the customer, making 10 $1 bills in their other hand, the clerk actually owes the customer $10. Thus, when the clerk gave the customer $20, they were giving the customer $10 too much.

This change-raising swindle is still perpetrated today in one form or another throughout the world. Here is yet another example of a change-raising con.

A stranger walks into a store. They hand a $20 bill to the clerk asking for change in the form of 3 $5 bills and 5 single dollar bills. The store clerk obliges and gives the stranger the relevant change. The stranger then walks toward the door intending to leave but returns and says that they have just recalled that they have several single $1 bills they wish to get rid of. They produce a small bundle of bills and say that there are 10 $1 bills in the bundle and ask to exchange them for a $10 bill. The clerk accepts the bundle of bills and simultaneously hands a $10 bill to the stranger. The clerk immediately counts the dollar bills on the store counter that were just handed to them and finds that there are just 9 bills in the bundle, and not 10. They mention this to the stranger, saying they have been left short $1. The stranger apologizes for the error and takes some bills from their wallet. They place a $1 bill on top of the 9 $1 bills on the counter, saying, "That makes 10," and then places 2 $5 bills on top of the pile of bills, adding, "and those 2 $5 bills make 20." The stranger then suggests that to make things simple, the clerk should just give them the original $20 bill that they (the stranger) gave the clerk earlier in exchange for the bills on the counter.

The clerk obliges. The stranger then leaves the store, having cheated the store out of $10.

Change-raising swindles are usually perpetrated by con artists posing as genuine customers in stores, bars, or gas stations. But sometimes the scammer may be operating as a merchant. On November 6, 2006, the BBC show *The Real Hustle*, which aired weekly in the United Kingdom for several years, illustrated a nice, short con. The scam involved one of the show's stars, a very attractive lady named Jessica-Jane Clement, playing the part of a lady selling ice pops on a public street. The whole scam was secretly filmed by hidden cameras.

A customer approaches Jessica and inquires about purchasing the ice pops. Jessica informs them that each one costs $1. The customer decides to purchase 2 and tenders $10 to Jessica. The vendor rummages through her change to obtain change for the customer. She puts the 2 ice pops to one side near her and says simultaneously, "A dollar each, so that's $2," and proceeds to give the customer their change by handing them $1 at a time saying slowly, "3, 4, 5, 6, 7, 8," and, at this point, Jessica hands the customer the 2 ice pops, saying, "and the pops make 10." Jessica then says to the customer: "Thank you very much. Have a lovely day. Thank you!"

The customer walks away, happy with their purchase, oblivious to the fact that the vendor has swindled them out of $2. (Presumably the customer discovers later that they only received $6 in change, instead of the $8 they were due, from the vendor.)

The game of Nim is an ancient 2-person mathematical game that con artists participate in to lure a sucker to part with their cash. The game is usually played with coins, matches, or playing cards on the counter of a bar.[7]

The game is usually played with 3 rows of coins, with 3 coins in the top row, 5 coins in the middle row, and 7 coins in the third row. However, the game can be played with any number of rows and with any number of coins in each row.

The rules of the game are simple. Each player takes alternate turns at removing 1 or more coins, provided they all come from the same horizontal row. The player taking the last coin wins. (Sometimes the game is played in reverse: the player taking the last coin loses. In the example here, the player taking the last coin wins.)

The hustler will usually introduce the game to unsuspecting customers at a bar. They will say that they like to take a gamble on each game of Nim and are therefore prepared to bet $5 per game at even money odds that they will win the game.

Of course, the customers at the bar do not realize—unless they know the secrets of the game—that they are almost certain to lose to the swindler. As the con artist sees things, they are laughing all the way to the bank if they can get suckers to participate in the con game.

Over a century ago mathematicians fully analyzed the game of Nim. Scam artists have also familiarized themselves with the game in order to profit from it. Hustlers have noted that the game (in either the normal form or reverse form) can always be won if, after making one of their moves, the con artist can leave 2 horizontal rows with more than 1 coin in each row, and the 2 rows containing the same number of coins; or if after making 1 of their moves, there is 1 coin in the top row, 2 coins in the second row and 3 coins in the third row. Also, in the layout of the coins given here, the hustler will know that if they have the first move, they can always win by taking 1 coin from the top row.

But what if the swindler has the second move? In that case the mark can win, but only by good luck, or if they know the mathematical secrets of the game.

Here's the mathematical analysis of the game. If, after making their move, a player is guaranteed to win the game *if* they play from that point onward in a rational manner, then the game is said to be in *safe* position. If, however, after making a move, the player is not guaranteed to win the game *if* they play rationally from that point onward, the position of the game is said to be *unsafe*.

To determine whether a move makes a game safe or unsafe, the con artist expresses the numbers in each row as the sum of the multiples of 2. If each column adds up to zero or an even number, then the position of the game is *safe*. If any column adds up to an odd number, the position of the game is *unsafe*.

In our example, the 3 rows consist of 3, 5, and 7 coins. When these numbers are expressed as the sum of powers of 2, we obtain the following situation:

		8	4	2	1
3	=			1	1
5	=		1	0	1
7	=		1	1	1
Totals			2	2	3

In our example here, the column at the far right adds up to 3, which is an odd number. Therefore, the position of the game is *unsafe*. A player can make the position safe by removing 1 coin from the top row. The con artist will know this, of course, and if they have the first move, they will take a coin from the top row. From that point onward the hustler cannot lose, *if* they play rationally. The position of the game after the hustler has made their first move is:

		8	4	2	1
2	=			1	0
5	=		1	0	1
7	=		1	1	1
Totals			2	2	2

Thus, the position of the game is now *safe*. *Provided the hustler plays in a rational manner from now on, they cannot lose.*

Suppose the con artist has the second move. Let us assume the mark on their first move removes 2 coins from the middle row. Then, after they remove the 2 coins, the position of the game is as follows:

		8	4	2	1
3	=			1	1
5	=			1	1
7	=		1	1	1
Totals			1	3	3

The position of the game is now *unsafe*.

The con artist can visualize the numerical analysis of the game almost instantly in their head. Thus, they know that the position of the game is now *unsafe*. To make the game *safe* they must remove all 7 coins from the bottom row. Once they do this, the position of the game then is:

		8	4	2	1
3	=			1	1
3	=			1	1
0	=		0	0	0
Totals				2	2

This position of the game is now *safe*.

The hustler has therefore engineered the game into a *safe* position again. All they need to do now is to play rationally from that point on and they are guaranteed to win the game.

The game of Nim is a great scam for the hustler. Don't play a game of Nim for money against a stranger. If a stranger offers you a game of Nim for money, there is a high probability that the stranger is a con artist. If they are, you will very likely lose money.

The following 2 scams are often perpetrated wherever the game of darts is played. This is mainly in bars throughout the United Kingdom and Ireland. But darts is also played widely in the United States. The game involves 2 players, who play against each other. They stand at a specific distance from the dartboard and, alternating turns, each throws a group of 3 darts. At the beginning of the game, each player starts with a total of 301 points, or sometimes 501 points. The object of the game is to reduce that total to zero in as few throws as possible.

To start scoring, a player must first shoot 1 of their 3 darts into a special section in the dartboard known as a "double segment." When that has been achieved, their scoring begins. Whatever score they achieve with the 3 darts is then subtracted from 301. As the game proceeds, each player is subtracting the amount they score from their previous total. The winning player is the one who first reduces their total to zero, but their last dart must end in the "doubles segment" of the board. In the parlance of dart players, one must start and finish with "a double."

According to the rules of the game, the center of the dartboard must be 5 feet and 8 inches above the floor. The players throwing the darts must stand behind a line that is 7 feet and 9.25 inches from the dartboard.

Witnessing a game of darts in a pub is usually enough to motivate a swindler into action. The con artist will approach a group of dart players and offer to play any of them under the following conditions: you can start with any odd number (this is usually 301 or 501) but must start and end with a "double." The con artist states that any score they get (after their first double is scored) will be retained as such. Thus, if the con artist scores 30, their score will be recorded as 30. But the con artist says that any score their opponent gets will be instantly doubled. Thus, if the grifter's opponent scores 15 with 1 dart, that score immediately counts as 30. If the mark scores 18 with 1 dart, that score is recorded as 36. And so on. The fraudster then bets $20, offering odds of 5 to 1 in favor of the challenger, that they will beat any opponent who is prepared to take them on.

If there is a good darts player among the customers in the bar, the bet will appear very attractive and, thus, at least one of the customers is likely to challenge the hustler in the hope of picking up a bundle of greenbacks. They agree to start the game, beginning with some odd total, usually 501.

But the hustler is laughing all the way to the bank on this bet! No matter how good the challenger is, they will not beat the con artist under the stated rules. Why? Well, you see, by the conditions laid down by the hustler, the challenger doubles their score each time they score. Thus, every score they obtain is doubled, and therefore becomes an *even* number. Thus, the challenger is constantly subtracting *even* numbers from an *odd* number.

For example, if the challenger scores 16, 21, and 24 with their first 3 darts, their scores are recorded as 32, 42, and 48. This is a total of 122, which is an *even* number. They must subtract this *even* number from 501, which leaves them with 379. This is an *odd* number. As the game proceeds the challenger (mark) most probably feels they are progressing well, as their scores are quickly accumulating, and they are therefore swiftly reducing the amount of points they began with and heading speedily toward zero.

However, it eventually dawns on the mark that they cannot finish the game on a "double," as they are required to do, in order to win the contest. To finish on a "double," one must have an *even* number of points *left*. Because they are constantly subtracting *even* numbers from *odd* numbers they will always be left with an *odd* number. The mark cannot reach an *even* number. Thus, they cannot finish on a "double"!

An example will clarify the matter. Suppose the mark is left with 27 to score. They have 3 darts to reduce this total of 27 to 0. The mark shoots for 1 and gets it. This would normally leave the mark with 26, or double 13 to get, to finish the game. But that 1 they have scored is now doubled, so it counts as 2. This leaves the mark with 25. They may decide with their second dart to shoot for 5. Assume they get the 5. That would normally leave the mark with 20, or double 10 to get, to finish the game. But that score of 5 is doubled to 10. This leaves them with 15 point. The mark may decide to aim for 3 with their third dart. Assume they get 3. That would normally leave them with 12, or double 6 to get, to finish the game. But the 3 they scored is doubled to 6, which leaves them with 9 to score the next time they throw their 3 darts. Sooner or later, the mark realizes they cannot get onto an *even* number. Therefore, they cannot shoot a "double." They cannot win! Their situation is hopeless! The hustler is laughing all the way to the bank.

The second proposition bet involving the game of darts goes as follows: a hustler will pin a $20 bill across the center of the dartboard. They will then bet anyone $20 at even money odds to throw 3 darts at the board, with the intention of trying to hit the bill with all 3 darts. Any challenger must put $20 in a glass. If they win the bet, they get back the $20 they have put in a glass and win the $20 bill pinned to the dartboard.

The conditions for the game are that the first dart is thrown from the normal starting line of 7 feet and 9.25 inches from the board. The second dart must be thrown from 8 feet, 9.25 inches from the board. The third dart must be thrown from 6 feet, 9.25 inches from the board. By varying the distances like this the con artist is ensuring that even experienced dart players will find it difficult to even hit the dart board with all 3 darts. Even extremely good players will fail to strike the $20 bill with 2 of their 3 darts under these conditions. The hustler will most probably walk away with the mark's $20 bill, plus their own bill that they had pinned to the dartboard.

One of the best proposition bets that I know of involving playing cards goes as follows: The hustler will ask you to examine and then shuffle a normal 52-card deck. They will then bet you $10 at even money odds that if you scan through the deck, you will find a queen lying next to a king or jack.[8]

On being presented with this bet, the mark will likely reason as follows: It seems very unlikely that a queen will be found lying next to a king or jack. Any 1 queen can occupy any of the 52 positions in a deck of cards. There are only 4 queens, 4 jacks, and 4 kings in a normal deck. It appears that a queen has many positions it can occupy besides being adjacent to a king or jack. Thus, it is unlikely that this stranger will win the bet.

The mark becomes convinced that the hustler has got the odds in this bet all wrong. Thus, it is likely that the mark will accept the bet. They both place $10 on a table beside them.

The mark examines the deck and finds it is a normal deck of 52 cards. They then thoroughly shuffle the deck and when they scan through the shuffled deck there—to their amazement—is a queen lying next to a king or a jack. The appearance of a queen lying adjacent to a king or a jack appears to have been against the odds.

This is a classic proposition bet. It may be hard to believe but the probability that a queen will be found adjacent to a king or jack in a shuffled deck is 75.085 percent!

In other words, if one were to thoroughly shuffle the deck and scan through the shuffled deck and this was repeated many times (say 100,000 times), 75.085 percent of the time you would find a queen lying next to a king or jack, as shown in figure 51.1. This is a totally surprising and counterintuitive result. I am grateful to Connor Stoyle for obtaining this result. I am also grateful to Drs. Lara Hawchar and Cónall Kelly, both from the School of Mathematical Sciences, University College, Cork, Ireland. All 3 of the above ran Monte Carlo simulations on their computers, which confirmed that the probability of a queen appearing next to a jack or a king in a shuffled deck is indeed very close to 75 percent. (Lara Hawchar's simulation stated that a queen was likely to be next to a king or a jack precisely 75.085 percent of the time.)

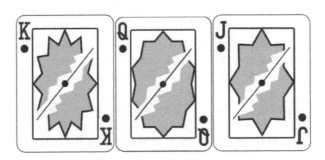

Figure 51.1 Face cards.

I will conclude the book by mentioning one more mathematical proposition bet.

You may be in a bar one evening when a stranger approaches and converses with you. The stranger may eventually bring the topic around to gambling. They eventually take from their wallet 26 business cards that all look alike. One letter of the English alphabet is printed on one side of each card. The other side of each card is blank. The stranger asks you to examine the cards to see that all 26 letters of the alphabet appear on the cards.

You examine the cards and are satisfied that each of the 26 cards does indeed contain 1 unique letter of the alphabet.

The stranger then asks you to thoroughly shuffle all 26 cards. You do so.

They will then ask you to count out 10 cards facedown and place them in a pile on a table beside you. Call that pile of 10 cards Pile A. They then ask you to place the other 16 cards facedown in a second pile beside the first pile. Let's call that second pile of 16 cards Pile B.

You again do as you are asked.

The stranger will now point out to you that neither of you know what letters are on the cards in either pile. They ask you to name any 2 letters of the alphabet and point out to you that you have a completely free choice in choosing these 2 letters.

You name the 2 letters of your choice.

The stranger will then say that they wish to try a little experiment. They are going to bet you $10, at 2 to 1 odds in your favor, that the 2 letters you have chosen, are to be found in Pile B.

Thus, if you accept the bet and win it, you will win $20, and receive your stake of $10 back. If, however, you lose the bet you will lose your $10 stake.

You are tempted at the thought of making easy money and left wondering if you should accept the bet.

What should you do?

As with all the previous proposition bets, do not accept this bet.

The number of ways 2 cards can be chosen from 10 is 10C2. This equals 45. Thus, there are 45 sets of 2 cards in Pile A, where each card contains a unique letter of the alphabet.

The number of ways 2 cards can be chosen from 16 is 16C2. This equals 120. Thus, in Pile B, there are 120 sets, each containing 2 cards, where each card contains a unique letter of the alphabet.

We see that there are 75 more ways of choosing 2 letters from 16 than there are of choosing 2 letters from 10. Thus, it is much more probable that any 2 letters you name will be from the 120 sets of 2 letters that have been selected from the 16 letters.

The odds of winning this bet are 120 to 45, which simplify to 8 to 3, in favor of the con artist. Saying the same thing another way, the probability of your winning the wager is 27.27+ percent. Thus, the probability that the con artist will win the bet is 72.72+ percent.

Even if the hustler gives you odds of 2 to 1, the bet is still a great money-maker—for the fraudster!

For every 100 bets, at $10 a round, and the hustler's giving the mark odds of 2 to 1, the con artist can expect to win about 72 $10, or $720, and lose about 28 $20, or $560. That is still a profit of $160 for the fraudster. That equates to an *average profit* of $1.60 per bet.

This bet is a great little earner for the hustler.

Don't make life easy for them by accepting the bet!

NOTES

1. Tom M. Apostol, "A Short History of Probability," homepages.wmich.edu/mackey/Teaching/A Short History of Probability.
2. Matthew M. Conroy, "A Collection of Dice Problems: Problem 8," Mad and Moonly, December 2, 2020, https://www.madandmoonly.com/doctormatt/mathematics/dice1.pdf.
3. "Getting One of Each Number (1–6) Rolling N Dice," mathforum.org/library/drmath/view.
4. Nick Trost, *Gambling Tricks with Cards* (Columbus, OH: Trik-Kard Specialties, 1975).
5. "List of Confidence Tricks: Fiddle Game," Wikipedia, https://en.wikipedia.org/wiki/List_of_confidence_tricks#Fiddle_game.
6. "List of Confidence Tricks: Change Raising," Wikipedia, https://en.wikipedia.org/wiki/List_of_confidence_tricks#Fiddle_game.
7. Martin Gardner, *Mathematical Puzzles and Diversions* (New York: Simon & Schuster, 1959), chapter 15: Nim and Tac Tix, 133–41.
8. Ed Collins, "Ed Answers Probability and Other Math-Related Questions," https://www.edcollins.com/math-questions.htm.

BIBLIOGRAPHY

Anderson, Harry. *Games You Can't Lose: A Guide for Suckers*. New York: Pocket Books, Simon & Schuster, 1989.

Dent, Paxton H. *Play Sucker and Pray*. El Paso, TX: Carl Hertzog, 1939.

Fields, Eddie. *A Life Among Secrets: The Uncommon Life and Adventures of Eddie Fields*. Seattle: Hermetic Press, 1992.

Fisher, John. *Never Give a Sucker an Even Break: Tricks and Bets You Can't Lose*. New York: Pantheon Books, 1976.

Gibson, Walter B. *The Bunco Book*. Las Vegas: Gamblers' Book Club, 1976.

———. *Carnival Gaffs*. Las Vegas: Gamblers' Book Club, 1976.

Gorham, Maurice. *Showmen and Suckers*. London: Percival Marshall, 1951.

Haigh, John. *Taking Chances: Winning with Probability*. New York: Oxford University Press, 1999.

Huff, Darrell. *How to Lie with Statistics*. London: Penguin Books, 1991.

———. *How to Take a Chance*. London: Victor Gollanz, 1960.

Jacoby, Oswald. *How to Figure the Odds*. Garden City, NY: Doubleday, 1947.

Lovell, Simon. *How to Cheat at Everything: A Con Man Reveals the Secrets of the Esoteric Trade of Cheating, Scams, and Hustles*. New York: Thunder's Mouth Press, 2003.

———. *Betcha! How to Win Free Drinks for Life*. Chicago: Magic Incorporated, 2014.

Maurer, David W. *The American Confidence Man*. Springfield, IL: Charles C. Thomas, 1974.

Prus, Robert C., and C. R. D. Sharper. *Road Hustler: The Career Contingencies of Professional Card and Dice Hustlers*. Lanham, MD: Lexington Books, 1977.

Rice, Charlie. *Challenge! Fun with Puzzles, Riddles, Word Games and Problems*. Kansas City, MO: Hallmark Cards, 1968.

Scarne, John. *10 Best Proposition Bets of America's Big Time Gamblers*. North Bergen, NJ: John Scarne Games, 1975.

Sullivan, Patrick. *Bets You Can't Lose*. Los Angeles: Price Stern Sloan, 1979.

———. *More Bets You Can't Lose*. Los Angeles: Price Stern Sloan, 1986.

Tucker, W. M. *The Change Raisers*. Greenville, SC: Self-published, c. 1930.

Weaver, Warren. *Lady Luck: The Theory of Probability*. New York: Dover, 1963.

Wilson, Paul. *The Art of the Con: How to Think Like a Real Hustler and Avoid Being Scammed*. Lanham, MD: Lyons Press, 2014.

Zenon, Paul. *100 Ways to Win a Tenner*. London: Carlton Books, 2003.